Introduction to Sports Biomechanics
Analysing Human Movement Patterns

Second edition

Roger Bartlett

Routledge
Taylor & Francis Group

LONDON AND NEW YORK

First edition published 1997
This edition first published 2007
by Routledge
2 Park Square, Milton Park, Abingdon, Oxon OX14 4RN

Simultaneously published in the USA and Canada
by Routledge
711 Third Avenue, New York, NY 10017 (8th Floor)

Routledge is an imprint of the Taylor & Francis Group, an informa business

© 1997, 2007 Roger Bartlett

Typeset in Adobe Garamond and Frutiger by
RefineCatch Limited, Bungay, Suffolk
Printed and bound in Great Britain by
TJ International Ltd, Padstow, Cornwall

British Library Cataloguing in Publication Data
A catalogue record for this book is available from the British Library

Library of Congress Cataloging in Publication Data
A catalog record for this book has been requested

ISBN10: 0–415–33993–6 (hbk)
ISBN10: 0–415–33994–4 (pbk)
ISBN10: 0–203–46202–5 (ebk)

ISBN13: 978–0–415–33993–3 (hbk)
ISBN13: 978–0–415–33994–0 (pbk)
ISBN13: 978–0–203–46202–7 (ebk)

Introduction to Sp[...]

Introduction to Sports Biomechanics: Analysing Human Movement Patterns provides a genuinely accessible and comprehensive guide to all of the biomechanics topics covered in an undergraduate sports and exercise science degree.

Now revised and in its second edition, *Introduction to Sports Biomechanics* is colour illustrated and full of visual aids to support the text. Every chapter contains cross-references to key terms and definitions from that chapter, learning objectives and summaries, study tasks to confirm and extend your understanding, and suggestions to further your reading.

Highly structured and with many student-friendly features the text covers:

- Movement Patterns – Exploring the Essence and Purpose of Movement Analysis
- Qualitative Analysis of Sports Movements
- Movement Patterns and the Geometry of Motion
- Quantitative Measurement and Analysis of Movement
- Forces and Torques – Causes of Movement
- The Human Body and the Anatomy of Movement

This edition of *Introduction to Sports Biomechanics* is supported by a website containing video clips, and offers sample data tables for comparison and analysis and multiple-choice questions to confirm your understanding of the material in each chapter.

This text is a must have for students of sport and exercise, human movement sciences, ergonomics, biomechanics and sports performance and coaching.

Roger Bartlett is Professor of Sports Biomechanics in the School of Physical Education, University of Otago, New Zealand. He is an Invited Fellow of the International Society of Biomechanics in Sports and European College of Sports Sciences, and an Honorary Fellow of the British Association of Sport and Exercise Sciences, of which he was Chairman from 1991–4. Roger is currently Editor of the journal *Sports Biomechanics*.

To the late James Hay, a source of great inspiration

Contents

Figures

Tables

Boxes

Preface

Why have I changed the cover name for this book from that of the first edition? Because after teaching, researching and consulting in sports biomechanics for over 30 years, my definition of sports biomechanics has become simply, 'the study and analysis of human movement patterns in sport'. This is a marked change from the first edition, the introduction to which began with the sentence: 'Sports biomechanics uses the scientific methods of mechanics to study the effects of various forces on the sports performer'. The change in focus – and structure and contents – of this book reflects an important change in sports biomechanics over the last decade. Most sports biomechanics textbooks, including the first edition of this one, have strongly reflected the mathematical, engineering or physics backgrounds of their authors and their predominant research culture. Hence, the mechanical focus that is evident, particularly in earlier texts, as well as a strong emphasis on quantitative analysis in sports biomechanics. In this early part of the third millennium, more students who graduate with a degree focused on sports biomechanics will go on to work as a movement analyst or performance analyst with sports organisations and client groups in exercise and health than will enrol for a research degree. The requirements on them will be to undertake mostly qualitative, rather than quantitative, analysis of movement. Indeed, I will often use the term 'movement analyst' instead of 'sports biomechanist' to reflect this shift from quantitative to qualitative analysis, and to broaden the term somewhat, as will be apparent later.

So, qualitative analysis is the main focus of the first three chapters of this new edition; however everything in these chapters is also relevant for quantitative movement analysts – you cannot be a good quantitative movement analyst without first being a good qualitative analyst. The last three chapters focus on quantitative analysis. Even here, there are notable changes from the first edition. First, I have removed sections that dealt with sports objects rather than the sports performer. This reflects the growth of sports engineering as the discipline that deals with the design and function of sports equipment and sports objects. Secondly, rather than the structure of the first edition – four chapters on fundamentals and four on measurement techniques – the measurement sections are now incorporated within Chapters 4 to 6 (and touched on in Chapter 2) and are covered only in the detail needed for undergraduate students. More advanced students wishing to probe deeper into measurement techniques and data processing will find the new text edited by Carl Payton and myself a source of more detailed information (*Biomechanical Evaluation of Movement in Sport and Exercise*, Routledge, 2007).

So what do sports biomechanists – or movement analysts – do? We study and analyse human movement patterns in sport to help people perform their chosen sporting activity better and to reduce the risk of injury. We also do it because it is so fascinating. Yes, it is fascinating, otherwise so many of my generation would not still be doing it. And it is intellectually challenging and personally gratifying – if you can contribute to reducing an athlete's injury risk or to improving his or her performance, it gives you a warm glow. Sounds exciting, doesn't it? Indeed it is – a wealth of fascination. So, let us begin our journey.

This edition is intended to be more reader-friendly than the first. Each chapter starts with an outline of learning outcomes, and knowledge assumed, which is cross-referenced mostly to other parts of the book. At the end of each chapter, a summary is provided of what was covered and eight study tasks are listed. Hints are given about how to go about each task, including referring to video clips, data tables and other material available on the book's website, which is, in itself, another important pedagogical resource. The website also includes PowerPoint slides for lecturers to use as a basis for their lectures, and multiple choice questions for students to self-test their learning progress. Further reading material is also recommended at the end of each chapter.

The production of any textbook relies on the cooperation of many people other than the author. I should like to acknowledge the invaluable, carefully considered comments of Dr Melanie Bussey on all the chapters of the book and, particularly, her glossaries of important terms in each chapter. All those who acted as models for the photographic illustrations are gratefully acknowledged: former colleagues of mine at Manchester Metropolitan University in the UK – Drs Vicky Goosey, Mike Lauder and Keith Tolfrey – and colleagues and students at the University of Otago in New Zealand – Dr Melanie Bussey, Neil Davis, Nick Flyger, Peter Lamb, Jo Trezise and Nigel Barrett – and Nigel's son Bradley; I thank Chris Sullivan for his help with some of the illustrations. I am also grateful to Raylene Bates for the photo sequence of javelin throwing, to Harold Connolly for the hammer throwing sequence, to Warren Frost for the one of bowling in cricket, and to Clara Soper for those of lawn bowling. I should not need to add that any errors in the book are entirely my responsibility.

Roger Bartlett, Dunedin, New Zealand

Introduction

The first three chapters of this book focus mainly on qualitative analysis of sports movements. Chapter 1 starts by outlining a novel approach to sports biomechanics and establishing that our focus in this chapter is the qualitative analysis of human movement patterns in sport. I will define movements in the sagittal plane and touch on those in the frontal and horizontal planes. We will then consider the constraints-led approach to studying human movements, and go on to look at examples of walking, running, jumping and throwing, including the subdivision of these fundamental movements into phases. In these movements, we will compare movement patterns between ages, sexes, footwear, inclines and tasks. The chapter concludes with a comparison of qualitative and quantitative analysis, looking at their background, uses, and strengths and weaknesses.

Chapter 2 considers how qualitative biomechanical analysis of movement is part of a multidisciplinary approach to movement analysis. We will look at several structured approaches to qualitative analysis of movement, all of which have, at their core, the identification of critical features of the movement studied. We will identify four stages in a structured approach to movement analysis, consider the main aspects of each stage and note that the value of each stage depends on how well the previous stages have been implemented. We will see that the most crucial step in the whole approach is how to identify the critical features of a movement, and we will look at several ways of doing this, none of which is foolproof. We will work through a detailed example of the best approach, using deterministic models, and consider the 'movement principles' approach and the role of phase analysis of movement.

Chapter 3 covers the principles of kinematics – the geometry of movement – which are important for the study of movement in sport and exercise. Our focus will be very strongly on movement patterns and their qualitative interpretation. Several other forms of movement pattern will be introduced, explained and explored – including stick figures, time-series graphs, angle–angle diagrams and phase planes. We will consider the types of motion and the model appropriate to each. The importance of being able to interpret graphical patterns of linear or angular displacement and to infer from these the geometry of the velocity and acceleration patterns will be stressed. We will look at two ways of assessing joint coordination using angle–angle diagrams and, through phase planes, relative phase, and we will briefly touch on the strengths and weaknesses of these two approaches. Finally, I present a cautionary tale of unreliable data as a warning to the analysis of data containing unacceptable measurement errors, providing a backdrop for the last three chapters.

Chapters 4 to 6 focus mainly on quantitative analysis of sports movements. Chapter 4 covers the use of videography in the study of sports movements, including the equipment and methods used. The necessary features of video equipment for recording movements in sport will be considered, along with the advantages and limitations of two- and three-dimensional recording of sports movements. I will outline the possible sources of error in recorded movement data and describe experimental procedures that would minimise recorded errors in two- and three-dimensional movements. The need for smoothing or filtering of kinematic data will be covered, and the ways of performing this will be touched on. I will also outline the requirement for accurate body segment inertia parameter data and how these can be obtained, and some aspects of error analysis. Projectile motion will be considered and equations presented to calculate the maximum vertical displacement, flight time, range and optimum projection angle of a simple projectile for specified values of the three projection parameters. Deviations of the optimal angle for the sports performer from the optimal projection angle will be explained. We will also look at the calculation of linear velocities and accelerations caused by rotation and conclude with a brief consideration of three-dimensional rotation.

Chapter 5 deals with linear 'kinetics', which are important for an understanding of human movement in sport and exercise. This includes the definition of force, the identification of the various external forces acting in sport and how they combine, and the laws of linear kinetics and related concepts, such as linear momentum. We will address how friction and traction influence movements in sport and exercise, including reducing and increasing friction and traction. Fluid dynamic forces will also be considered and I will outline the importance of lift and drag forces on both the performer and on objects for which the fluid dynamics can impact on a player's movements. We will emphasise both qualitative and quantitative aspects of force–time graphs. The segmentation method for calculating the position of the whole body centre of mass of the sports performer will be explained. The vitally important topic of rotational kinetics will be covered, including the laws of rotational kinetics and related concepts such as angular momentum and the ways in which rotation is generated and controlled in sports motions. The use of force plates in sports biomechanics will be covered, including the equipment and methods used, and the processing of force plate data. We will also consider the important measurement characteristics required for a force plate in sports biomechanics. The procedures for calibrating a force plate will be outlined, along with those used to record forces in practice. The different ways in which force plate data can be processed to obtain other movement variables will be covered. The value of contact pressure measurements in the study of sports movements will also be considered. Some examples will be provided of the ways in which pressure data can be presented to aid analysis of sports movements.

Chapter 6 focuses on the anatomical principles that relate to movement in sport and exercise. This includes consideration of the planes and axes of movement and the principal movements in those planes. The functions of the skeleton, the types of bone, bone fracture and typical surface features of bone will be covered. We will then look briefly at the tissue structures involved in the joints of the body, joint stability and mobility, and the identification of the features and classes of synovial joints. The features and structure of skeletal muscles will be considered along with the ways in which muscles are structurally and functionally classified, the types and mechanics of muscular contraction, how tension is produced in muscle and how the total force exerted by a muscle can be resolved into components depending on the angle of pull. The use of electromyography (EMG) in the study of muscle activity in sports bio-mechanics will be considered, including the equipment and methods used, and the processing of EMG data. Consideration will be given to why the electromyogram is important in sports biomechanics and why the recorded EMG differs from the physio-logical EMG. We will cover the relevant recommendations of SENIAM and the equipment used in recording the EMG, along with the main characteristics of an EMG amplifier. The processing of the raw EMG signal will be considered in terms of its time domain descriptors and the EMG power spectrum and the measures used to define it. We will conclude by examining how isokinetic dynamometry can be used to record the net muscle torque at a joint.

1 Movement patterns – the essence of sports biomechanics

Knowledge assumed
**Familiarity with human
movement in sport
Ability to undertake simple
analysis of videos of sports
movements**

INTRODUCTION

What were my reasons for choosing the title of this book and the name of this chapter? Well, after teaching, researching and consulting in sports biomechanics for over 30 years, my definition of the term has become, quite simply, 'the study and analysis of human movement patterns in sport'.

Nothing about 'the scientific methods of mechanics' or 'the effects of various forces' or 'Newton's laws' or vectors or . . .? No, nothing like that – just 'the study and analysis of human movement patterns in sport'. Sounds exciting, doesn't it? Indeed it is – a wealth of fascination. So, let us begin our journey.

Having offered my definition of sports biomechanics, it becomes obvious what sports biomechanists do – we study and analyse human movement patterns in sport. But why do we do it? Well, the usual reasons are:

- To help people perform their chosen sporting activity better. We should note here that this does not just apply to the elite athlete but to any sportsperson who wants to improve his or her performance.
- To help reduce the risk of injury.

From a pedagogical perspective, we might add:

- To educate new generations of sports biomechanists, coaches and teachers.

And, from a personal viewpoint:

- Because it is so fascinating. Yes, it is fascinating, otherwise so many of my generation would not still be doing it. It is also intellectually challenging and personally gratifying – if you can contribute to reducing an athlete's injury risk or to improving his or her performance, it gives you a warm glow.

Most sports biomechanics textbooks, including the first edition of this one, have strongly reflected the mathematical, engineering or physics backgrounds of their authors and their predominant research culture. Hence, the mechanical focus that is evident, particularly in earlier texts, as well as a strong emphasis on quantitative analysis in sports biomechanics. However, over the last decade or so, the 'real world' of sport and exercise outside of academia has generated – from coaches, athletes and other practitioners – an increasing demand for good qualitative movement analysts. Indeed, I will often use the term 'movement analyst' instead of 'sports biomechanist' to reflect this shift from quantitative to qualitative analysis, and I will broaden the term somewhat, as will be apparent later. So, qualitative analysis is our main focus in this chapter –

BOX 1.1 LEARNING OUTCOMES

After reading this chapter you should be able to:

- think enthusiastically about analysing movement patterns in sport
- understand the fundamentals of defining joint movements anatomically
- appreciate the differences – and the similarities – between qualitative and quantitative analysis of sports movements
- describe, from video observation or pictorial sequences, some simple sport and exercise movements, such as walking, running, jumping and throwing
- appreciate why breaking these movements down into phases can help simplify their description and later analysis
- be familiar with finding supplementary information – particularly videos – on the book's website
- feel enthusiastic about progressing to Chapters 2 and 3.

and the next two. However, everything in these chapters is also relevant for quantitative movement analysts – you cannot be a good quantitative movement analyst without first being a good qualitative analyst.

DEFINING HUMAN MOVEMENTS

In this section, we look at how we can define human movements, something to which we will return in more detail in Chapter 6. To specify unambiguously the movements of the human body in sport, exercise and other activities, we need to use an appropriate scientific terminology. Terms such as 'bending knees' and 'raising arms' are acceptable in everyday language, including when communicating with sport practitioners, but 'raising arms' is ambiguous and we should strive for precision. 'Bending knees' is often thought to be scientifically unacceptable – a view with which I profoundly disagree as I consider that simplicity is always preferable, particularly in communications with non-scientists. We need to start by establishing the planes in which these movements occur and the axes about which they take place, along with the body postures from which we define these movements. These planes, axes and postures are summarised in Box 1.2.

BOX 1.2 PLANES AND AXES OF MOVEMENT AND POSTURES FROM WHICH MOVEMENTS ARE DEFINED

Various terms are used to describe the three mutually perpendicular intersecting planes in which many, although not all, joint movements occur. The common point of intersection of these three planes is most conveniently defined as either the centre of the joint being studied or the centre of mass of the whole human body. In the latter case, the planes are known as cardinal planes – the sagittal, frontal and horizontal planes – as depicted in Figure 1.1 and described below.

Movements at the joints of the human musculoskeletal system are mainly rotational and take place about a line perpendicular to the plane in which they occur. This line is known as an axis of rotation. Three axes – the sagittal, frontal and vertical (longitudinal) – can be defined by the intersection of pairs of the planes of movement, as in Figure 1.1. The main movements about these three axes for a particular joint are flexion and extension about the frontal axis, abduction and adduction about the sagittal axis, and medial and lateral (internal and external) rotation about the vertical (longitudinal) axes.

- The sagittal plane is a vertical plane passing from the rear (posterior) to the front (anterior), dividing the body into left and right halves, as in Figure 1.1(a). It is also known as the anteroposterior plane. Most sport and exercise movements that are almost two-dimensional, such as running and long jumping, take place in this plane.

- The frontal plane is also vertical and passes from left to right, dividing the body into posterior and anterior halves, as in Figure 1.1(b). It is also known as the coronal or the mediolateral plane.
- The horizontal plane divides the body into top (superior) and bottom (inferior) halves, as in Figure 1.1(c). It is also known as the transverse plane.
- The sagittal axis (Figure 1.1(b)) passes horizontally from posterior to anterior and is formed by the intersection of the sagittal and horizontal planes.
- The frontal axis (Figure 1.1(a)) passes horizontally from left to right and is formed by the intersection of the frontal and horizontal planes.
- The vertical or longitudinal axis (Figure 1.1(c)) passes vertically from inferior to superior and is formed by the intersection of the sagittal and frontal planes.

The movements of body segments are usually defined from the fundamental (Figure 1.2(a)) or anatomical (Figure 1.2(b)) reference postures – or positions – demonstrated by the athlete in Figure 1.2. Note that the fundamental position is similar to a 'stand to attention', as is the anatomical position, except that the palms face forwards in the latter.

Figure 1.1 Cardinal planes and axes of movement: (a) sagittal plane and frontal axis; (b) frontal plane and sagittal axis; (c) horizontal plane and vertical axis.

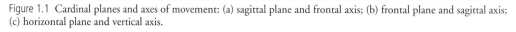
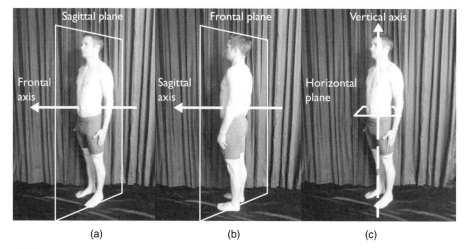

(a) (b) (c)

Figure 1.2 Reference postures (positions): (a) fundamental and (b) anatomical.

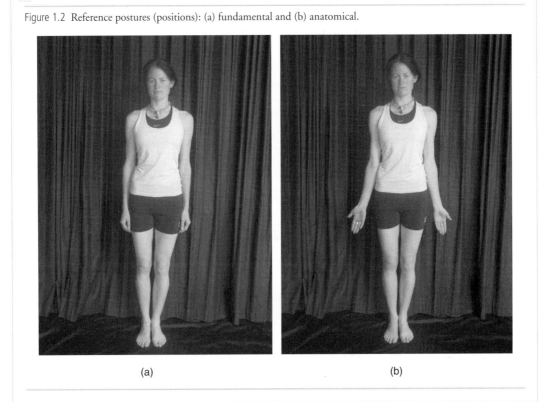

(a)   (b)

By and large, this chapter focuses on movements in the sagittal plane about the frontal (or mediolateral) axis of rotation (Figure 1.1(a)). Consider viewing a person side-on, as in Figure 1.3; he bends his elbow and then straightens it. We call these movements flexion and extension, respectively, and they take place in the sagittal plane around the frontal axis of rotation. Flexion is generally a bending movement, with the body segment – in the case of the elbow, the forearm – moving forwards. When the knee flexes, the calf moves backwards. The movements at the ankle joint are called plantar flexion when the foot moves downwards towards the rear of the calf, and dorsiflexion when the foot moves upwards towards the front of the calf.

The movement of the whole arm about the shoulder joint from the anatomical reference position is called flexion, and its return to that position is called extension; the continuation of extension beyond the anatomical reference position is called hyperextension. The same terminology is used to define movements in the sagittal plane for the thigh about the hip joint. These arm and thigh movements are usually defined with respect to the trunk.

Sports biomechanists normally use the convention that the fully extended position of most joints is 180°; when most joints flex, this angle decreases. Clinical bio-mechanists tend to use an alternative convention in which a fully extended joint is 0°, so that flexion increases the joint angle. We will use the former convention throughout

Figure 1.3 Movement of the forearm about the elbow joint in the sagittal plane – flexion and extension.

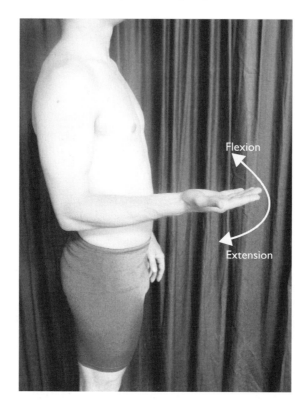

this book. Because the examples of movement patterns that we will study in this chapter are mainly in the sagittal plane, we will leave formal consideration of movements in the other two planes until Chapter 6. The main ones are shown in Box 1.3.

BOX 1.3 MAIN MOVEMENTS IN OTHER PLANES

Movements in the frontal plane about a sagittal axis are usually called abduction away from the body and adduction back towards the body, as in Figure 1.4. For some joints, such as the elbow and knee, these movements are not possible, or are very restricted.

Movements in the horizontal plane about a vertical axis are usually called medial (or internal) and lateral (or external) rotation for the limbs, as in Figure 1.5, and rotation to the right or to the left for the trunk. The movements of the whole arm forwards from a 90° abducted position are horizontal flexion in a forwards direction and horizontal extension in a backwards direction, as in Figure 1.6.

Figure 1.4 Abduction and adduction of the arm about the shoulder joint and the thigh about the hip joint.

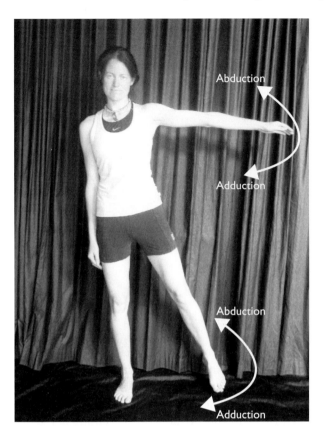

Figure 1.5 Medial (internal) and lateral (external) rotation of the arm about the shoulder joint.

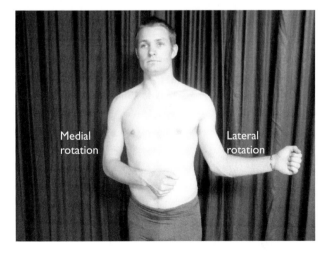

Figure 1.6 Horizontal flexion and extension of the abducted arm about the shoulder joint.

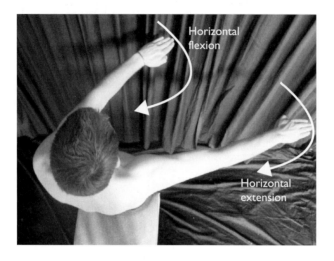

SOME FUNDAMENTAL MOVEMENTS

The website for this book contains many video clips of various people performing some movements that are fundamental to sport and exercise. These people include the young and the old, male and female, who are shown walking, running, jumping and throwing in various conditions. These include: locomotion on a level and inclined treadmill and overground; vertical and broad jumping; underarm, sidearm and overarm throwing; in different footwear and clothing; and with and without skin markers to identify centres of rotation of joints. An in-depth study of these videos is recommended to all readers. The video sequences shown in the figures below have been extracted from these clips using the qualitative analysis package siliconCOACH™ (siliconCOACH Ltd, Dunedin, New Zealand; http://www.siliconcoach.com).

When analysing any human movement, ask yourself, 'What are the "constraints" on this movement?' The constraints can be related to the sports task, the environment or the organism. This 'constraints-led' approach serves as a very strong basis from which to develop an understanding of why we observe particular movement patterns. In the video examples and the sequences in the figures below, an environmental constraint might be 'overground' or 'treadmill' (although this might also be seen as a task constraint). Jumping vertically to achieve maximum height is clearly a task constraint. Organismic constraints are, basically, biomechanical; they relate to a given individual's body characteristics, which affect their movement responses to the task and environmental constraints. These biomechanical constraints will be affected, among many other things, by genetic make-up, age, biological sex, fitness, injury record and stage of rehabilitation, and pathological conditions. Not surprisingly, the movement patterns observed when one individual performs a specific sports task will rarely be identical to those of another person; indeed, the movement patterns from repetitions of that task by the same individual will also vary – this becomes more obvious when we quantitatively analyse those movements, but can be seen qualitatively in many patterns of movement, as in Chapter 3. These variable responses, often known as movement variability, can and do affect the way that movement analysts look at sports movements. The qualitative descriptions in the following sections will not, therefore, apply to every adult, but will apply to many so-called 'normals'. The developmental patterns of maturing children up to a certain age show notable differences from those for an adult, as in Figures 1.12 and 1.18 below.

A first step in the analysis of a complex motor skill is often to establish the phases into which the movement can be divided for analysis. For example, the division of a throwing movement into separate, but linked, phases is useful because of the sheer complexity of many throwing techniques. The phases of the movement should be selected so that they have a biomechanically distinct role in the overall movement, which is different from that of preceding and succeeding phases. Each phase then has a clearly defined biomechanical function and easily identified phase boundaries, often called key events. Although phase analysis can help the understanding of movement patterns, the essential feature of all sports movements is their wholeness; this should always be borne in mind when undertaking any phase analysis of a movement pattern.

Walking

Walking is a cyclic activity in which one stride follows another in a continuous pattern. We define a walking stride as being from touchdown of one foot to the next touchdown of the same foot, or from toe-off to toe-off. In walking, there is a single-support phase, when one foot is on the ground, and a double-support phase, when both are. The single-support phase starts with toe-off of one foot and the double-support phase starts with touchdown of the same foot. The duration of the single-support phase is about four times that of the double-support phase. Alternatively, we can consider each leg separately. Each leg then has a stance and support phase, with similar functions to those in running (see pages 15–23). In normal walking at a person's preferred speed, the stance phase for one leg occupies about 60% of the whole cycle and the swing phase around 40% (see, for example, Figure 1.7). In normal walking, the average durations of stance and swing will be very similar for the left and right sides. In pathological gait, there may be a pronounced difference between the two sides, leading to arrhythmic gait patterns.

The book's website contains many video clips of side and rear views of people walking. These illustrate differences between males and females, between young and older adults and young children, between overground and treadmill locomotion and at different speeds and treadmill inclines, and with various types of footwear. Figures 1.7 to 1.12 contain still images from a selection of these video clips. Observing these figures, you should note, in general, the following patterns of flexion and extension of the hip, knee and ankle; you should also study the video sequences on the book's website to become familiar with identifying these movements on video. The hip flexes

Figure 1.7 Young female walking overground at her preferred speed in trainers. Top left: left foot touchdown (0 s); top right: right foot toe-off (0.12 s); middle left: left foot mid-stance; middle right: right foot touchdown (0.52 s); bottom left: left foot toe-off (0.64 s); bottom right: right foot mid-stance.

during the swing phase and then begins to extend just before touchdown; extension continues until the heel rises just before toe-off. The hip then starts to flex for the next swing phase, roughly when the other foot touches down. The knee is normally slightly flexed at touchdown and this flexion continues, although not necessarily in slow walking. Some, but not much, extension follows before the knee starts to flex sharply immediately after the heel rises; this flexion continues through toe-off until about

Figure 1.8 Same young female as in Figure 1.7 walking on a level treadmill at her preferred speed in trainers. Top left: right foot touchdown (0 s); top right: left foot toe-off (0.14 s); middle left: right foot mid-stance; middle right: left foot touchdown (0.52 s); bottom left: right foot toe-off (0.64 s); bottom right: left foot mid-stance.

halfway though the swing, when the knee extends again, before flexing slightly just before touchdown. The ankle is roughly in a neutral position at touchdown, as in the reference positions of Figure 1.2. The ankle then plantar flexes until the whole foot is on the ground, when dorsiflexion starts; this continues until the other leg touches down. Plantar flexion then follows almost to toe-off, just before which the ankle dorsiflexes quickly to allow the foot to clear the ground as it swings forwards. As you

Figure 1.9 Older male walking on a level treadmill at his preferred speed in bowling shoes. Top left: left foot touch-down (0 s); top right: right foot toe-off (0.14 s); middle left: left foot mid-stance; middle right: right foot touchdown (0.50 s); bottom left: left foot toe-off (0.64 s); bottom right: right foot mid-stance.

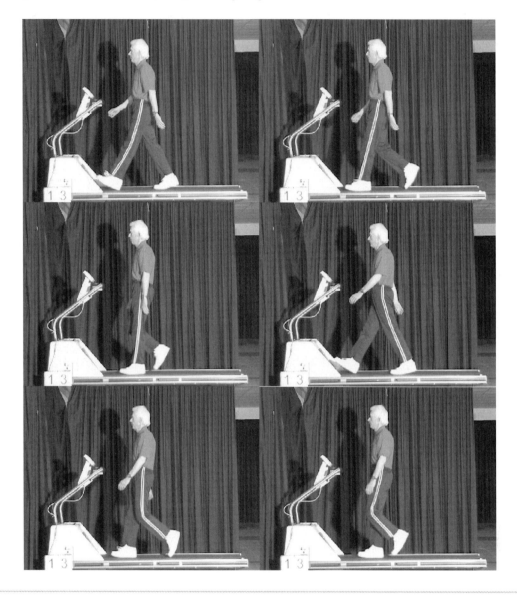

should note from Figures 1.7 to 1.12 and from the video clips on the book's website, this sequence of movements varies somewhat from person to person (see also, for example, Figure 3.11(a)), with the shoes worn, the surface inclination, the walking speed, and between overground and treadmill walking. The movement pattern for a child walking (Figure 1.12) is very different from that of an adult.

So, what would we seek to observe as movement analysts looking at walking patterns

Figure 1.10 Another young female walking on a level treadmill at her preferred speed in high-heeled shoes. Top left: left foot touchdown (0 s); top right: right foot toe-off (0.14 s); middle left: left foot mid-stance; middle right: right foot touchdown (0.52 s); bottom left: left foot toe-off (0.64 s); bottom right: right foot mid-stance.

(we would then be functioning as gait analysts)? Differences from this normal pattern, for one, but also right–left side differences, variations across strides, how joint and contralateral limb movements are coordinated, and how external factors, such as changed task or environmental constraints, affect the gait pattern. Video sequences, as in Figures 1.7 to 1.12, are not necessarily the best movement pattern representation for these purposes, as we shall see in Chapter 3.

Figure 1.11 Young male walking on a 20% inclined treadmill at his preferred speed in work shoes. Top left: left foot touchdown (0 s); top right: right foot toe-off (0.16 s); middle left: left foot mid-stance; middle right: right foot touchdown (0.52 s); bottom left: left foot toe-off (0.68 s); bottom right: right foot mid-stance.

Figure 1.12 Three-year-old boy walking overground. Top left: left foot touchdown (0 s); top right: right foot toe-off (0.06 s); middle left: left foot mid-stance; middle right: right foot touchdown (0.38 s); bottom left: left foot toe-off (0.44 s); bottom right: right foot mid-stance.

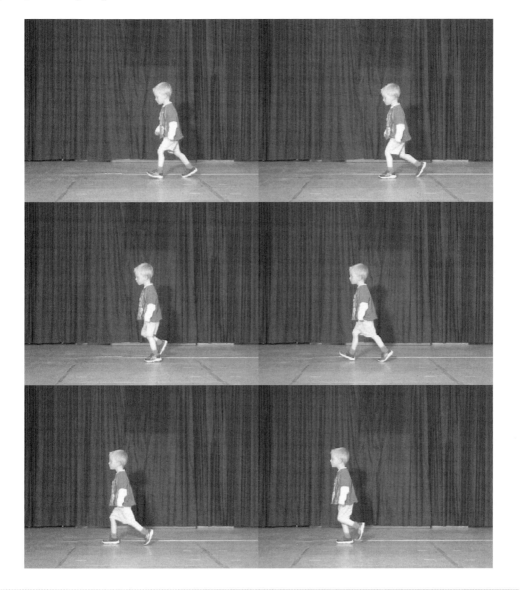

Running

Running, like walking, is a cyclic activity; one running stride follows another in a continuous pattern. We define a running stride as being from touchdown of one foot to the next touchdown of the same foot, or from toe-off to toe-off. Unlike walking (see

Figure 1.13 Young female running at her preferred speed in trainers. Top left: left foot toe-off (0 s); top right: right foot touchdown (0.18 s); middle left: right foot mid-stance; middle right: right foot toe-off (0.42 s); bottom left: left foot touchdown (0.58 s); bottom right: left foot mid-stance.

pages 9–15), running can basically be divided into a support phase, when one foot is on the ground, and a recovery phase, in which both feet are off the ground. The runner can only apply force to the ground for propulsion during the support phase, which defines that phase's main biomechanical function and provides the key events that indicate the

Figure 1.14 Another young female running at her preferred speed in dress shoes. Top left: right foot toe-off (0 s); top right: left foot touchdown (0.14 s); middle left: left foot mid-stance; middle right: left foot toe-off (0.38 s); bottom left: right foot touchdown (0.54 s); bottom right: right foot mid-stance.

start of the phase, touchdown (or foot strike), and its end, toe-off. The recovery phase starts at toe-off and ends at touchdown; at this stage, we will consider its function to be to prepare the leg for the next touchdown. In slow running, or jogging, the recovery phase will be very short; it will then increase with running speed.

As for walking, the book's website also contains many side- and rear-view video

Figure 1.15 Young male running at his preferred speed in casual shoes. Top left: left foot toe-off (0 s); top right: right foot touchdown (0.12 s); middle left: right foot mid-stance; middle right: right foot toe-off (0.36 s); bottom left: left foot touchdown (0.48 s); bottom right: left foot mid-stance.

clips of people running. These illustrate differences between males and females, between young and older adults and young children, between overground and treadmill locomotion, at different speeds, and with various types of footwear. Figures 1.13 to 1.19 contain still images from a selection of these video clips. Observing these figures, you should note, in general, the following patterns of flexion and extension of the hip, knee and ankle; you should also study the video sequences on the book's website to become

Figure 1.16 Older male running at his preferred speed in normal trainers. Top left: right foot toe-off (0 s); top right: left foot touchdown (0.12 s); middle left: left foot mid-stance; middle right: left foot toe-off (0.38 s); bottom left: right foot touchdown (0.50 s); bottom right: right foot mid-stance.

familiar with identifying these movements on video. The hip continues to extend early in the swing phase, roughly until maximum knee flexion, after which it flexes then begins to extend just before touchdown; extension continues until toe-off. The knee is normally slightly flexed at touchdown and this flexion continues, depending on running speed, to absorb shock, until the hip is roughly over the ankle. Knee extension then proceeds until toe-off, soon after which the knee flexes as the hip continues to extend. The knee starts to extend while the hip is flexing and continues to extend

Figure 1.17 Older male running at his preferred speed in MBT™ trainers. Top left: left foot toe-off (0 s); top right: right foot touchdown (0.12 s); middle left: right foot mid-stance; middle right: right foot toe-off (0.34 s); bottom left: left foot touchdown (0.44 s); bottom right: left foot mid-stance.

almost until touchdown, just before which the knee might flex slightly. The ankle movements (see also, for example, Figure 3.13(b)) vary depending on whether the runner lands on the forefoot or rear foot. The ankle is roughly in a neutral position at touchdown, as in the reference positions of Figure 1.2. For a rear foot runner, in particular, the ankle then plantar flexes slightly until the whole foot is on the ground; dorsiflexion then occurs until mid-stance. The ankle plantar flexes from mid-stance

Figure 1.18 Three-year-old boy running at his preferred speed. Top left: left foot toe-off (0 s); top right: right foot touchdown (0.08 s); middle left: right foot mid-stance; middle right: right foot toe-off (0.24 s); bottom left: left foot touchdown (0.30 s); bottom right: left foot mid-stance.

until toe-off, as the whole support leg lengthens. The ankle then dorsiflexes to a neutral position in the swing phase and plantar flexes slightly just before touchdown. As you should note from Figures 1.13 to 1.19, and from the video clips on the book's website, this sequence of movements varies somewhat from person to person (see also, for example, Figure 3.11(b)), with the shoes worn, with running speed, and between overground and treadmill running. The movement pattern for a child running (Figure 1.18) is very different from that of an adult.

Figure 1.19 Young male sprinting in spikes. Top left: left foot toe-off (0 s); top right: right foot touchdown (0.12 s); middle left: right foot mid-stance; middle right: right foot toe-off (0.24 s); bottom left: left foot touchdown (0.38 s); bottom right: left foot mid-stance.

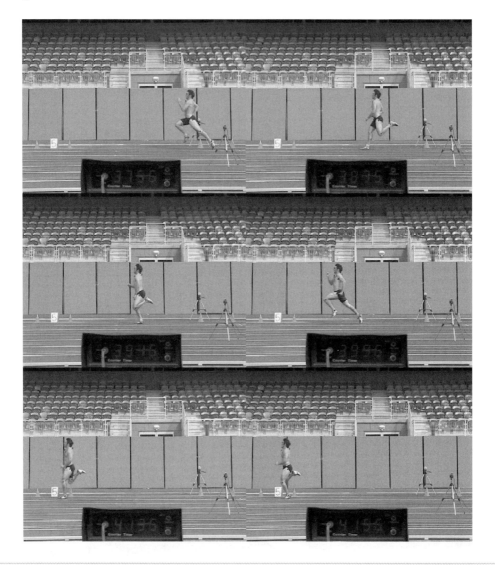

So, what would we seek to observe as movement analysts looking at running patterns? Differences from this normal pattern, certainly, but also right–left side differences, variations across strides, and how joint movements are coordinated within a limb as well as between legs and with the arm movements. We might also want to look at how external factors, such as changed task or environmental constraints, affect the running pattern. As we also noted for walking, video sequences (as in

Figures 1.13 to 1.19), are not necessarily the best movement pattern representation for these purposes.

Jumping

Jumps, as well as throws, are often described as 'ballistic' movements – movements initiated by muscle activity in one muscle group, continued in a 'coasting' period with no muscle activation, and terminated by deceleration by the opposite muscle group or by passive tissue structures, such as ligaments. Many ballistic sports movements can be subdivided biomechanically into three phases: preparation, action and recovery. Each of these phases has specific biomechanical functions. In countermovement jumps from a standing position, such as those in Figures 1.20 to 1.25, the preparation is a lowering phase, which puts the body into an advantageous position for the action (raising) phase and stores elastic energy in the eccentrically contracting (lengthening) muscles. The action phase has a synchronised rather than sequential structure, with all leg joints extending or plantar flexing together. The recovery phase involves both the time in the air and a controlled landing, the latter through eccentric contraction of the leg muscles.

Figure 1.20 Standing countermovement vertical jump with hands on hips. Top left: starting position; top right: lowest point; bottom left: take-off; bottom right: peak of jump.

Figure 1.21 Standing countermovement vertical jump with normal arm action. Top left: starting position; top right: top point of arm swing; middle left: lowest point; middle right: take-off; bottom: peak of jump.

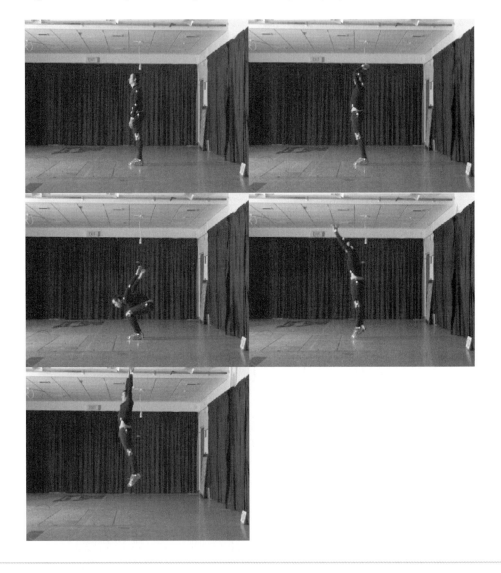

Jumps that involve a run-up, such as the long or high jump, or that have a more complex structure, such as the triple jump, benefit from being divided into more than three phases. In jumps with arm movements, the coordination of the arm actions with those of the legs is very important to performance.

Figure 1.22 Standing countermovement vertical jump with 'model' arm action. Top left: starting position; top right: lowest point; middle left: take-off; middle right: peak of jump.

The standing vertical jump

The standing vertical jump looks simple. The extensor muscles of the hips and knees and the plantar flexors of the ankle contract eccentrically to allow the knees and hips to flex and the ankles to dorsiflex simultaneously in the preparation phase. The action phase involves the simultaneous extension of the hips and knees and plantar flexion of the ankles through shortening (concentric) contraction of the muscles that extend or plantar flex these joints and drive the body vertically upwards. This sequence is evident in Figures 1.20 to 1.23. The main difference between the countermovement jump with no arm action in Figure 1.20 and that with a free arm action in Figure 1.21 is that the arm actions in the latter jump, if properly coordinated with those of the legs, will enhance performance of the jump. You should compare Figure 1.21, in which the jumper used his normal arm action, with the simpler arm action in Figure 1.22, based on a simple biomechanical 'model', and the uncoordinated arm action in Figure 1.23. The jumper performs as well with the model action as with his normal action, part of which is nearly identical to the model. However, the arm action of Figure 1.23, which is roughly the reverse of the model action, causes a marked decline in jump performance. In the model and normal jumps, the arm and leg movements are well

Figure 1.23 Standing countermovement vertical jump with abnormal arm action. Top left: starting position; top right: lowest point; middle left: take-off; middle right: peak of jump.

coordinated, unlike in the abnormal jump, in which the arm and leg movements are poorly coordinated.

In a standing vertical jump, we would first seek to observe coordination of the movements within and between the legs, and of the leg movements with those of the arms. The standing vertical jump is often used as a field test of leg power, so the movement needs to be fast and powerful, as well as coordinated, to result in a successful – and high – jump.

The standing broad, or long, jump

The sequence of movements and the principles of the standing long – or broad – jump are very similar to those of the standing vertical jump. However, as the task is now to jump as far as possible horizontally, the jumper needs to partition effort between the vertical and horizontal aspects of the jump, mainly through forward lean – this somewhat complicates the task. As in the standing vertical jump, the coordinated swing of the arms improves performance, as can be seen by comparing the jump without (Figure 1.24) and with (Figure 1.25) an arm swing. Coordination of all limb actions is again critically important. We would also look for a take-off angle of 35–45° as an

Figure 1.24 Standing countermovement broad, or long, jump with hands on hips. Top left: starting position; top right: lowest point; bottom: take-off.

indicator of how well the jumper had partitioned effort between the horizontal and vertical components of the jumps. We could do this by trying to observe the difference between the height of the jumper's centre of mass – indicated roughly by the height of the hips – at take-off and at landing. The higher the take-off height above the landing height, the smaller the take-off angle should be. If the take-off and landing heights are equal, the optimum angle would be 45° (see also Chapter 4, page 145).

Figure 1.25 Standing countermovement broad, or long, jump with normal arm action. Top left: starting position; top right: arms at highest point; bottom left: lowest point; bottom right: take-off.

Throwing

This section focuses on the principles of those sports or events in which the participant throws, passes, bowls or shoots an object from the hand or, in the case of lacrosse, from an implement. Some, or all, of these principles relate to: throws from a circle – hammer and discus throws, shot put; crossover skills – javelin throw and cricket bowling; pitching in baseball and softball; shooting and passing movements in basketball, netball, handball, water polo and lacrosse; throwing-to skills – baseball, cricket, soccer, rugby, American and other variants of football; underarm bowling; and dart throwing. Some of these are used as examples in this section. As with other ballistic sports movements, many throws can be subdivided biomechanically into three phases: preparation, action and recovery. Each of these phases has specific biomechanical functions. The later phases depend upon the previous phase or phases. In a basic throw, such as those in Figures 1.26 to 1.28, the preparation phase puts the body into an advantageous position for the action phase and increases the acceleration path of the object to be thrown. In skilled throwers, the action phase demonstrates a sequential action of muscles as segments are recruited into the movement pattern at the correct time. The recovery

Figure 1.26 Underarm throw – female bowling a 'drive'. Top left: starting position; top right: end of backswing; bottom left: delivery; bottom right: follow-through.

phase involves the controlled deceleration of the movement by eccentric contraction of the appropriate muscles. Throws that have a more complex structure, such as the hammer throw (Figure 1.29), or that involve a run-up, such as javelin throwing (Figure 1.30) or cricket bowling (Figure 1.31), benefit from being divided into more than three phases (see also Appendix 2.2 for a phase breakdown of the javelin throw).

Throwing movements are often classified as underarm, overarm or sidearm. The last two of these can be viewed as diagonal movement patterns, in which trunk lateral flexion, the trunk bending sideways, is mainly responsible for determining whether one of these throws is overarm or sidearm. In the overarm pattern, the trunk laterally flexes away from the throwing arm, in a sidearm pattern the trunk laterally flexes towards that arm.

The goal of a throwing movement will generally be distance, accuracy or some combination of the two, acting as a task constraint. The goal is important in determining which of the movement principles discussed in detail in Chapter 2 are more, and which are less, applicable. Some movement analysts distinguish between throw-like movements for distance, in which segmental rotations occur sequentially, and push-like movements for accuracy, in which segmental rotations occur simultaneously. However, few throws in sport have no accuracy requirements. Even those, such as javelin,

Figure 1.27 Underarm throw – female bowling a 'draw'. Top left: starting position; top right: end backswing; bottom left: delivery; bottom right: follow-through.

discus and hammer throwing and shot putting, in which the distance of the throw is predominant, have to land in a specified area and have rules that constrain the throwing technique. In throws for distance, the release speed – and therefore the force applied to the thrown object – is crucial, a theme to which we will return in Chapter 4.

In some throws, the objective is not to achieve maximal distance: instead, it may be accuracy or minimal time in the air. The latter is particularly important in throws from the outfield in baseball and to the wicketkeeper in cricket. In such throws, the release speed, height and angle need to be such that the flight time is minimised within the accuracy and distance constraints of the throw. In accuracy-dominated skills, such as dart throwing and some passes and set shots in basketball, the release of the object needs to achieve accuracy within the distance constraints of the skill. The interaction of speed and accuracy in these skills is often expressed as the speed–accuracy trade-off. This can be seen, for example, in a basketball shot in which the shooter has to release the ball with both speed and accuracy to pass through the basket.

Figure 1.28 Underarm throw – young male bowling a 'draw'. Top left: starting position; top right: end of backswing; bottom left: delivery; bottom right: follow-through.

Underarm throws

Underarm throws with one arm are characterised by the shoulder action, which is predominantly flexion, often from a hyperextended position above the horizontal, as for the fast drive shot in lawn bowling (Figure 1.26) but not for the slower draw shot (Figures 1.27 and 1.28). In the preparation phase, the weight transfers to the rear foot and the front foot steps forwards; this step is often longer for skilled throwers. Weight transfers onto the front foot during the action phase, as the pelvis and trunk rotate to the left (for a right-handed thrower). The elbow extends during the action phase and, at release, the throwing arm is parallel to, or slightly in front of, the line of the trunk. Curling and softball pitching are underarm throws, as are tenpin and various other bowling actions used, for example, in lawn bowling (Figures 1.26 to 1.28), tenpin bowling and skittles. You should study carefully the sequences in Figures 1.26 to 1.28 and the video clips on the book's website to observe differences between individuals and tasks in underarm throwing, or bowling, patterns (see Study task 7). Video clips of these and other underarm throws, such as the softball pitch and rugby spiral pass (the latter of which is a two-handed throw), are also available on the book's website.

Figure 1.29 Sidearm throw – the hammer throw. Top left: entry to turns; top right: first turn; middle left: second turn; middle right: third turn; bottom left: fourth turn; bottom right: release.

Sidearm throws

Sidearm throws are sometimes considered to differ from underarm and overarm throws, mainly by restricted action at the shoulder joint. The dominant movement is rotation of the pelvis and trunk with the arm abducted (see Box 1.2) to a position near the horizontal. Unlike the other two throwing patterns, in which the movements are mainly in the sagittal, or a diagonal, plane, frontal plane movements dominate in sidearm throws. The discus throw and some shots in handball are of this type. The hammer

Figure 1.30 Overarm throw – javelin throw. Top left: withdrawal of javelin; top right: start of crossover; middle left: crossover; middle right: start of delivery stride; bottom left: left foot landing; bottom right: release.

throw (Figure 1.29) is probably best characterised as a sidearm throw rather than an underarm throw.

Overarm throws

Overarm throws are normally characterised by external rotation (see Box 1.2) of the upper arm in the preparation phase and by its internal rotation in the action phase.

Figure 1.31 Overarm throw – bowling in cricket. Top left: approach; top right: start of bound; middle left: back foot landing; middle right: front foot landing; bottom left: release; bottom right: follow-through.

These movements are among the fastest joint rotations in the human body. Many of the other joint movements are similar to those of the underarm throw. The sequence of movements in the preparation phase of a baseball pitch, for example, include (for a right-handed pitcher), pelvic and trunk rotation to the right, horizontal extension and lateral rotation at the shoulder, elbow flexion and wrist hyperextension. These movements are followed, sequentially, by their anatomical opposite at each of the joints mentioned plus internal rotation of the forearm, also known as pronation.

Baseball pitching, javelin throwing (Figure 1.30), throwing from the outfield in cricket and passing in American football are classic examples of one-arm overarm throws. The mass (inertia) and dimensions of the thrown object – plus the size of the target area and the rules of the particular sport – are constraints on the movement pattern of any throw. Bowling in cricket (Figure 1.31) differs from other overarm throwing patterns, as the rules restrict elbow straightening (extension) during the latter part of the delivery stride. The predominant action at the shoulder is, therefore, circumduction – a combination of shoulder flexion, extension, abduction and adduction. The soccer throw-in uses a two-handed overarm throwing pattern. The shot put combines overarm throwing with a pushing movement, because of the event's rules and the mass of the shot. Basketball shooting uses various modifications to the overarm throwing pattern, depending on the rules of the game and the circumstances and position of the shot – including release speed and accuracy requirements. Passing in basketball, in which accuracy is also crucial, varies from the overarm patterns of the overhead and baseball passes to the highly modified pushing action of the chest pass. In dart throwing, the dominant requirement for accuracy restricts movements in the action phase to elbow extension with some shoulder flexion–abduction. You should study carefully the video clips of the different individuals on the book's website to observe differences in their overarm throwing patterns (see Study task 7).

MOVEMENT PATTERNS

Most of you (readers of this book) will be undergraduate students in the earlier stages of your career. You will be familiar with human movement patterns from sport – when viewed live, or as a performer, coach or spectator – whether these are movement patterns of individuals or of teams as a whole. An example for an individual sport can be presented as a sequence of still video frames, as in Figures 1.7 to 1.31; most packages for qualitative video analysis make it easy to observe, and to compare, such movement patterns.

Video recordings, still video sequences, and player tracking patterns in games are probably the most complex representations of sports movements that you will come across. It is only your familiarity with sports videos that enables you to understand such patterns – watch a video of a game or sporting activity for which you do not know the rules (environmental and task constraints), and the complexity of video representations of movement patterns becomes obvious. This is true not only for the movements of the segments of the body of one performer, which sports biomechanists generally focus on, but also for the movement patterns of the players as a team. Sequences of still video frames are rarely used in analysing player movements and interactions in team games, such as rugby, netball and soccer, or in individual vs individual games, such as squash or table tennis. To understand why, imagine tracking (using the Global Positioning System, for example), just a single point on each player in one extended squash rally or, worst still, for each player in a soccer team for just 10 minutes of play. The resulting movement patterns would not be easy to analyse at first sight. Such movement patterns in games will not be considered further in this book – sports biomechanists, to date, have rarely been involved in analysing such movement patterns.

To appreciate why I say that video recordings are complex, did you find it easy to follow all the flexion and extension descriptions for walking and running in the previous section? Could you easily perceive within-leg and between-leg coordination patterns in walking, or arm and leg coordination patterns in running, using the sequences above or videos from the book's website? If your answers to these questions are a resounding 'YES', then you are already a talented qualitative movement analyst! Many of us struggle at times to extract what we want from video or from selected video picture sequences; for one thing they contain so much information that is irrelevant to the patterns the movement analyst wishes to observe. So, what alternative representations of a movement are available, not only to the quantitative analyst but also to the qualitative analyst? We will answer this important question in Chapter 3.

COMPARISON OF QUALITATIVE AND QUANTITATIVE MOVEMENT ANALYSIS

Sports biomechanists use two main approaches to analysing human movement patterns in sport – qualitative and quantitative analysis. The previous section focused on qualitative analysis. A third approach fits somewhere between the two and is often known as semi-quantitative analysis. These approaches will be developed and explained more fully in later chapters, but here I give a bullet-pointed outline of each, focusing on the two main approaches, including why they are used and by whom, as well as some advantages and drawbacks of each.

Qualitative analysis

What do we use for this?

- Video recording or observation.
- Other movement pattern representations, such as graphs (see Chapter 3), focusing on their patterns, not their quantification.
- Qualitative analysis software packages, such as siliconCOACH™.

Who uses this?

- Teachers, coaches, athletes, physiotherapists, gait analysts, and judges of 'artistic' sports, such as ice dance and gymnastics.
- 'Performance analysts' working with athletes and others.
- Movement coordination researchers (this one might surprise you, but it shouldn't once you have read Chapter 3).

Why is it used?

- To differentiate between individuals or teams.
- To improve movement or performance, as in gait analysis and video analysis.
- To provide qualitative feedback.

Semi-quantitative analysis

What do we use for this?

- Mostly as for qualitative analysis plus some simple measurements such as:
  - joint ranges of motion
  - durations of sub-phases of the movement, such as the stance and support phases in running, and their ratios to the overall movement time
  - distances, such as stride length
  - joint angles at key times, such as knee angle at take-off for a jump
  - notation – goals scored, passes, etc.

Who uses this?

- Pretty much as for qualitative analysis excluding, perhaps, teachers.

Why is it used?

- Pretty much as for qualitative analysis, but when comparisons are more important.
- Scaling aspects of a performance; for example, poor to excellent on a 1–5 point scale.

Quantitative analysis

What do we use for this?

- Image-based motion analysis, mostly using video (see Chapter 4) or automatic marker-tracking systems plus, when occasion demands, electromyography (see Chapter 6) and force or pressure plates or insoles (see Chapter 5).
- Statistical modelling of technique or of movement patterns in games, artificial intelligence, computer simulation modelling.
- Quantitative movement analysis and notational analysis software packages.

Who uses this?

- Mainly researchers.

Why is this used?

- To aid performance comparisons.
- To predict injury risk.
- To provide quantitative feedback.

Qualitative vs quantitative analysis

- Qualitative analysis describes and analyses movements non-numerically, by seeing movements as 'patterns', while quantitative analysis describes and analyses movement numerically.
- Quantitative analysis can sometimes appear more objective because of its 'data'; however the accuracy and reliability of such data can be very suspect, particularly when obtained in competition.
- Qualitative analysis is often more strongly rooted in a structured and multi-disciplinary approach, whereas quantitative analysis can appear to lack a theoretical grounding and to be data-driven.

Background to qualitative analysis

To be objective and scientific, qualitative analysis needs to use a structured approach, moving from preparation, through observation, diagnosis–evaluation, to intervention (and review) – this approach will be explained fully in Chapter 2. From the outset, the movement analyst should involve the coach, or whoever commissioned the analysis, in a 'needs analysis', and should keep the coach in the loop at all stages. Qualitative analysis requires applying basic biomechanical principles to the movement. We need to know what to observe; coaches have important knowledge and contributions to make here too.

Qualitative analysts need an excellent grasp of the techniques – or movement interactions – in a specific sport or exercise; coaches have great depth and breadth of that knowledge. Deterministic models (see Chapter 2) can give a theoretical basis to the analysis, which can otherwise become discursive. This modelling approach can be represented graphically so as to be coach-friendly.

Good-quality digital video cameras are needed, with adequate frame rates and shutter speeds. This equipment is familiar to coaches and extra equipment is rarely necessary. Qualitative analysis should uncover the major faults in an unsuccessful performance by an individual or a team; it is the approach actually used by most coaches and teachers.

Strengths and weaknesses of qualitative analysis

Strengths

- No expensive equipment (digital video cameras).
- Field-based not laboratory-based, which enhances ecological validity.
- When done properly, it is highly systematic.
- Movement patterns speak far more loudly than numbers – remember the cliché, a picture is worth a thousand words.
- Coach-friendly.

Weaknesses

- Apparent lack of 'data' (but is this really such a weakness?).
- Need for considerable knowledge of movement by analysts.
- Reliability and objectivity are questionable and often difficult to assess; observer bias.

Background to quantitative analysis

Mathematical models based on biophysical laws can give a sound theoretical basis to the analysis, which can otherwise become data-driven; most of these models are too far removed from coaching to be of practical use. Good quantitative analysts need a sound grasp of techniques or movement interactions involved in a specific activity, as do good qualitative analysts. However, not all quantitative analysis follows this principle, which might make much of their work dubious in a practical context.

A quantitative analyst needs to decide upfront the measurement techniques and methods to obtain the information required. Careful attention should be paid to what to measure, research design, data analysis, validity and reliability.

Strengths and weaknesses of quantitative analysis

Strengths

- Lots of biomechanical data (but is this really a strength?).
- Reliability and objectivity can be easily assessed, even if they rarely are.

Weaknesses

- Expensive equipment and software; user requires technical skills.
- Often laboratory- and not field-based, which reduces ecological validity.
- Apparent lack of a theoretical basis.
- When done badly, which it often is, it is highly non-systematic.

- Need for careful data management, as there's so much information available.
- Not coach-friendly.

SUMMARY

We started this chapter by outlining a novel approach to sports biomechanics and establishing that our focus in this chapter would be the qualitative analysis of human movement patterns in sport. We defined movements in the sagittal plane and touched on those in the frontal and horizontal planes. We then considered the constraints-led approach to studying human movements, and went on to look at examples of walking, running, jumping and throwing, including the subdivision of these fundamental movements into phases. In these movements, we compared movement patterns between ages, sexes, footwear, inclines and tasks. We then compared qualitative and quantitative analysis, looking at their background, uses, and strengths and weaknesses.

STUDY TASKS

1 Name and sketch the movements at the hip, knee and ankle in the sagittal plane, as observed, for example, in walking, running and jumping.
   Hint: You may wish to reread the section on 'Defining human movements' (pages 3–6) and watch several video clips of walking and running on the book's website before undertaking this task.

2 Name, and illustrate, the movements about the shoulder in all three planes as might be observed, for example, in overarm throwing.
   Hint: You may wish to reread the section on 'Defining human movements' (pages 3–6) and the subsection on 'Overarm throws' (pages 33–5), and watch several video clips of throwing on the book's website before undertaking this task.

3 Outline the phases into which running and walking are most simply divided. Give one important role of each phase for each activity. What distinguishes walking from running?
   Hint: You may wish to reread the subsections on 'Walking' (pages 9–15) and 'Running' (pages 15–23) and watch several video clips of walking and running on the book's website before undertaking this task.

4 Download a walk-to-run transition sequence from the book's website. For each of the five speeds in the sequence calculate, by counting frames and division:
   (i) The ratios of the durations of the single-support to double-support phases and of both these phases to the overall stride time in walking.
   (ii) The ratios of the no-support to the single-support phases and of both these phases to the overall stride time in running.

Explain the changes in these ratios through the sequence from slow walking to fast running.

Hint: You may wish to reread the subsections on 'Walking' (pages 9–15) and 'Running' (pages 15–23) before undertaking this task.

5   Download the video sequences for the four standing vertical jumps and two standing long jumps for the young male from the book's website. Try to explain why the height and distance jumped are affected by the use of the arms.

Hint: You may wish to reread the subsection on 'Jumping' (pages 23–8) before undertaking this task.

6   Outline the three phases into which a throw is usually divided. What is the main function of each phase?

Hint: You may wish to reread the subsection on 'Throwing' (pages 28–35) before undertaking this task.

7   Download a series of video sequences from the book's website of various underarm, sidearm or overarm throwing movements by different people. Note and try to explain the differences between them.

Hint: You may wish to reread the subsection on 'Throwing' (pages 28–35) before undertaking this task.

8   Outline the advantages and disadvantages of qualitative and quantitative movement analysis and explain in what circumstances one would be preferred over the other.

Hint: You may wish to reread the section on 'Comparison of qualitative and quantitative movement analysis' (pages 36–40) before undertaking this task.

You should also answer the multiple choice questions for Chapter 1 on the book's website.

GLOSSARY OF IMPORTANT TERMS (compiled by Dr Melanie Bussey)

Axes The imaginary lines of a reference system along which position is measured.

Axis of rotation An imaginary line about which a body or segment rotates.

Ballistic Rapid movements initiated by muscular contraction but continued by momentum.

Countermovement A movement made in the direction opposite to that of the desired direction of motion – as in a countermovement jump.

Pathological movement An exceptionally (or awkwardly or inconveniently) and atypical example of movement usually linked to an underlying anatomical or physiological cause, such as injury or disease.

Plane of motion A two-dimensional plane running through an object. Motion occurs in the plane or parallel to it. Motion in the plane is often called planar.

Reference system A system of coordinates used to locate a point in space.

Reliability The consistency of a set of measurements or measuring instrument.

Spatial Refers to a set of planes and **axes** defined in three-dimensional space.

FURTHER READING

Hay, J.C. (1993) *The Biomechanics of Sports Techniques*, Englewood Cliffs, NJ: Prentice Hall. This well-regarded book by the late Dr James Hay was the first biomechanics text to influence my approach to the discipline some 30 years ago. He takes a typically mechanistic approach to biomechanics, but it is one of the easiest such texts for a student to follow. Read the sports chapters that interest you, as these are the greatest strength of the book – has any other biomechanics author had such knowledge and insight into so many sports movements?

2 Qualitative analysis of sports movements

Knowledge assumed
Ability to undertake simple
analysis of videos of sports
movements (Chapter 1)
Movements in sagittal plane
and main movements in
frontal and horizontal planes
(Chapter 1)
Fundamental movements,
such as running, jumping and
throwing (Chapter 1)

INTRODUCTION

In the previous chapter, we considered human motion in sport and exercise as 'patterns of movement', and introduced key aspects of the qualitative and quantitative analysis of human movement patterns in sport. This chapter is designed to provide a contextual structure for carrying out qualitative movement analysis. The approach adopted here is based on previous books on qualitative analysis of human movement (see Further Reading on page 76 for more details) and many years of practical experience. This approach is very similar to that recommended by some professional agencies that represent sports biomechanists, or by agencies that hire sports biomechanists, such as the New Zealand Academy of Sport. Although the approach outlined here is used more by qualitative than quantitative analysts, it could – and should – be adopted by the latter group to provide a structure for their work.

BOX 2.1 LEARNING OUTCOMES

After reading this chapter you should be able to:

- understand how qualitative biomechanical analysis fits within the multidisciplinary framework of qualitative movement analysis
- plan and undertake a qualitative video analysis of a sports technique of your choice
- develop a critical insight into qualitative biomechanical analysis of movement in sport and exercise
- appreciate the need for a structured approach to qualitative movement analysis
- outline the principles of deterministic modelling and perform a qualitative analysis of a sports skill in detail, using a hierarchical model
- appreciate how use of a deterministic model can avoid some of the pitfalls of qualitative analysis
- understand the roles, within qualitative analysis, of phase analysis of movements and the movement principles approach.

A STRUCTURED ANALYSIS FRAMEWORK

The approach outlined in this section, and developed further in the five sections that follow, focuses on the systematic observation and introspective judgement of the quality of human movement to provide the best intervention to improve performance. This approach is necessarily interdisciplinary and integrated, involving motor development, motor learning and biomechanics, together with some aspects of physiology and psychology. The now defunct Performance Analysis Steering Group of the British

Olympic Association, which I was privileged to chair in the early years of the millennium, recognised this interdisciplinarity through the make-up of the group, which included notational analysts, sports biomechanists and motor skills specialists, together with coaches and performers.

The various approaches used by movement analysts have focused on biomechanics, motor development or pedagogy, and have sometimes been cross-disciplinary. Previous work has included development approaches, for example looking at whole-body developmental sequences, as in the four stages of acquiring throwing skills, or adopting a 'movement-component approach', focusing on the legs, the arms or the trunk. More recent developments have included logical decision trees, as in Figure 2.1.

The focus has varied in the various pedagogical approaches. Sometimes the observer has been recommended to attend to the temporal phases and spatial aspects of the movement. Other approaches have integrated various disciplines, have considered the pre-observation, observation and post-observation stages of analysis, and introduced the concept of critical features – those that contribute most to successful performance of the skill.

The various biomechanical approaches have typically identified the critical features of a skill using 'biomechanical principles' (better called 'movement principles'). These approaches include POSSUM – the Purpose-Observation System of Studying and Understanding Movement. In this approach, the movement is classified by its purpose, as in Figure 2.2(a), which is associated with 'observable dimensions' of the movement that the observer evaluates. In the example of Figure 2.2(a), the focus is a projectile – the whole body of a sports performer or an object, such as a shot. If the purpose is height, then the release or take-off is vertical; if the purpose is horizontal distance (range), the release is around 45° (but see Chapter 4 for further consideration of this point); if the purpose is speed, the release is nearly horizontal. The focus of the observer can be on the whole body or on specific body segments. This approach was extended around ten core concepts of 'kinesiology', as in Figure 2.2(b). Other biomechanical approaches have also tended to be based on a list of movement principles, for example Figure 2.2(c). It is instructive to compare such sets of principles, such as those

Figure 2.1 Simplified logical decision tree approach to qualitative classification of fast bowling technique.

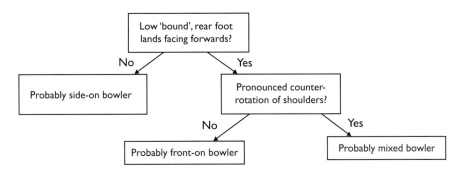

Figure 2.2 'Principles' approach to qualitative analysis (adapted from Knudson and Morrison, 2002): (a) classifying movement by its purpose; (b) core concepts of kinesiology; (c) movement principles approach; (d) comprehensive approach to qualitative analysis (for outline of details of each stage, see Box 2.2).

(a) Purpose   Observation

     Projection   Initial Path

              Height   Vertical

              Range   ~ 45°

              Speed   ~ Horizontal

(b)   Range of motion

                 Speed of motion

                 Number of segments

                 Balance

                 Coordination

                 Compactness

                 Extension at release

                 Projection path

                 Nature of segments

                 Spin

(c) Summation of joint torques

     Continuity of joint torques

     Impulse

     Reaction

     Equilibrium

     Summation and continuity of body segment velocities

     Generation of angular momentum

     Conservation of angular momentum

     Manipulation of moment of inertia

     Manipulation of body-segment angular momentum

(d)

Preparation Stage

Observation Stage

(repeat as necessary)

Evaluation and Diagnosis Stage

Intervention (and Review) Stage

of Figures 2.2(b) and (c), and to note that they do not seem to correspond. This led several movement analysts, me included, to develop three subclasses of principles (see Appendix 2.1). The first of these – universal principles – apply to all sports movements. The second – partially general principles – apply to groups of sports movements, such as those in which the task constraints demand a focus on speed generation or accuracy. Finally come principles applying to a specific skill, such as javelin throwing (specific principles).

The 'deterministic' or 'hierarchical' modelling approach, to which we will return later in the chapter, incorporates movement principles within a more structured framework, which should reduce the need to memorise lists of principles.

Nearly all of these approaches focus on qualitative analysis based on identifying errors in the movement and how to correct them. Unfortunately, whichever approach has been used, the tendency has been to focus on instantaneous events, such as a leg, arm or trunk angle at release of an implement. Alas, such 'discrete parameters' often tell us little about the overall movement, the distinctive features of which are its wholeness, and its coordination or lack of it in novices. One of the aims of this book is to help to rectify this lack of focus on movement wholeness and coordination. The most convincing approach to a structured qualitative analysis of sports movements, in my view, is that of Knudson and Morrison (2002; see Further Reading, page 76), which I have overviewed and extended in the next five sections, and which is represented diagrammatically in Figure 2.2(d); this approach is summarised in Box 2.2 and elaborated on in the next four sections.

BOX 2.2 STAGES IN A STRUCTURED APPROACH TO ANALYSIS OF HUMAN MOVEMENT IN SPORT

Stage 1 – Preparation

- Conducting a 'needs analysis' with the people commissioning the study to ascertain what they want from it.
- Gathering knowledge of activity and performers.
- Establishing critical features of the movement and, possibly, their range of correctness; this moves into semi-quantitative analysis (Chapter 1).
- Developing a systematic observation strategy for stage 2.
- Deciding on other qualitative presentations of movement patterns to be used (Chapter 3).
- Knowledge of relevant characteristics of performers.
- Knowledge of effective instruction, including cue words and phrases or task sheets, for stage 4.

Stage 2 – Observation

- Implementing the systematic observation strategy developed in stage 1.
- Gathering information about movement from the senses and from video recordings.
- Focus of observation, for example on phases of movement.

- Where to observe – controlled environment or ecologically valid one.
- From where to observe movement (vantage points) – including consideration of other qualitative patterns (Chapter 3).
- Number of observations – there is no such thing as a representative trial.

Stage 3 – Evaluation and diagnosis

- Evaluation of strengths and weaknesses of the performance.
- Use of other qualitative movement patterns in addition to video images (Chapter 3).
- Validity and reliability – generally poor to moderate for a single analyst and poor across analysts; increased training of analyst and analysing more trials both help.
- Issues:
  - variability of movement
  - movement errors
  - critical features vs ideal form
  - analyst bias.
- Select the best intervention to improve performance; this involves judgement of causes of poor performance.
- Issues:
  - lack of theoretical basis (not necessarily)
  - prioritising intervention – several approaches based on features of the movement that:
    - ❖ relate to previous actions
    - ❖ promise greatest improvement in performance or reduction in injury risk
    - ❖ proceed in order of difficulty of changing, from easiest to hardest
    - ❖ are in a correct sequence
    - ❖ work from the base of support up.

Stage 4 – Intervention

- Emphasises feedback to performers to improve technique and performance.
- Main review of the overall qualitative analysis process in the context of the needs analysis in stage 1 – this does not preclude reviews at other stages of the process, as implied by the solid arrows in Figure 2.2(d).
- Many issues arise about how, when and where to provide feedback.
- Also raises issues about practice, which need to address motor control models.
- Also raises issues about technique – or skills – training, and other aspects of training.

PREPARATION STAGE – KNOWING WHAT AND HOW TO OBSERVE

As can be seen from Box 2.2, the preparation stage involves much gathering of knowledge, including the undertaking of a needs analysis, identifying critical features of the performance, and preparing for later stages in the analysis process. We will focus here mainly on the first of these – gathering knowledge. The identification of critical features of the performance is so important that I have devoted a large section to this aspect of the preparation stage later in the chapter.

The gathering of relevant knowledge is dynamic and ongoing. A successful movement analyst needs knowledge, first and foremost, of the activity or movement, from which he or she will then develop the critical features of performance. Secondly, knowledge is needed of the performers; this includes the needs of the performers, and coaches or therapists, which should be identified in the 'needs analysis'. Although the preparation for later stages of the qualitative analysis process takes place, at least in gathering relevant knowledge, in the preparation stage, these will also be dealt with later. Developing a systematic observation strategy is covered in the next section, and developing a feedback strategy is dealt with on pages 56–8. In Chapter 3, we will discuss which qualitative representations of movement patterns to use in the evaluation and diagnosis stage, in addition to video analysis.

Your knowledge of the activity, as a movement analyst, should draw on many sport and exercise science disciplines. For example, as a primary Physical Education teacher, you would source knowledge mainly from the discipline of motor development: a secondary Physical Education teacher, by contrast, would focus more on an analysis of individual skills and techniques using, primarily, biomechanics. As a movement analyst working with novices, motor learning and practice would be major sources of information for you. On the other hand, a movement analyst working with good club-standard performers would probably focus on a biomechanically-derived identification of critical features, and a movement analyst working with elite performers would concentrate on the critical features at that standard, and might use a more quantitative approach.

In all of our work as movement analysts, whether qualitative or quantitative, we should always seek to adhere to 'evidence-based' practice, which raises the question as to what evidence we gather and from where. A movement analyst has, in general, access to various sources of knowledge about the sports activity being studied. Some issues arise in using these sources, including the fragmentary nature of some sources and weighing conflicting evidence from various sources. Experience also influences success in using source material, and helps to deal with anecdotal evidence and, with care, personal bias. The gathering of valid knowledge of the activity under consideration is invaluable if done systematically, and one needs to keep practising developing critical features based on the knowledge gathered. A warning here is appropriate – although the Internet is a fruitful source of information, in general there is little, if any, quality control over what appears there. There are exceptions to this warning, particularly peer-reviewed websites such as the Coaches' Information Services site (http://coachesinfo.com/) run by the International Society of Biomechanics in Sports (ISBS; http://www.isbs.org). Valid information is best sourced from such expert opinion, which can also be found in professional journals such as the *Sport and Exercise Scientist* (British Association of Sport and Exercise Sciences; http://www.bases.org.uk) and sport-specific coaching journals (such as *Swimming Technique*, now an integral part of *Swimming World Magazine*; http://www.swimmingworldmagazine.com). Many of these sources are accessible through the Internet. The performers and their support staff included in any 'real-world' study are also a potential source of knowledge about their sport, as may be other coaches and performers involved; not all of their knowledge will normally be evidence-based, so care is needed in using it. Problems associated with synthesising all of this knowledge include conflicts of opinion, a reliance on the 'elite-athlete template' (i.e. what the most successful do must also be right for others) and incorrect notions about critical features.

Scientific research should provide the most valid and accurate sources of information. Movement analysts need some research training, however, to interpret research findings: applied BSc or MSc degrees should provide such training, while a research-focused PhD may not. The best sources of relevant, applied research are applied sports science research journals, such as *Sports Biomechanics*, published on behalf of the ISBS, and the best coaching journals, such as *New Studies in Athletics*. Sports-specific scientific review papers draw together knowledge from many sources and provide a valuable source of information for movement analysts, providing the reviews have an applied rather than a fundamental research focus. The *Journal of Sports Sciences* has been a fruitful source for such review papers. The major problem with scientific research as a source of information for the qualitative movement analyst might be called the validity conflict between internal (research) validity and ecological (real-world) validity.

It is not sufficient just to gather knowledge of the activity; it must also be theoretically focused and practically synthesised. Adopting a 'fundamental movement pattern' approach is now seen as flawed, because of its over-reliance on the motor program concept of cognitive motor control. The constraints-led approach, introduced briefly in Chapter 1, considers the movement 'space' (the set of all possible solutions to the specific movement task) as constrained by the task, environment and organism; this is the approach of ecological motor control, which is still evolving. The critical features approach, adopted below, is the most widely used by movement analysts from a sports biomechanics perspective. The analyst needs to keep practising this practical approach, whose points are widely used in teaching and coaching. The movement criterion might be injury risk, movement effectiveness – defined as achievement of the movement goal – or efficiency, the economical use of metabolic energy. Analysts often specify a range of correctness of critical features, and this range must be observable. One common error is not focusing sufficiently on devising cue words for use in correcting technique errors; error correction should be seen as the responsibility of the movement analysis team, which includes the coach and the movement analyst, not the coach alone.

Relevant knowledge of the performers will include their age, sex and standard of performance; physical abilities, such as fitness, strength and flexibility; injury status and history; and cognitive development, which relates to the feedback to be provided in the intervention stage. Also relevant here is knowledge of the particular activity as related to a specific performer, which may require knowledge from motor development and motor learning. An extremely important knowledge source is the 'needs' of the performers and their coaches or therapists; to address this properly requires a 'needs analysis' led by the movement analyst, based on the foregoing points and knowledge of the sports activity. This needs analysis (which in the real world must include a project costing to deliver what is needed) then has to be approved by your 'clients', before leading into the acquisition of other relevant knowledge, being translated into a systematic observation strategy, driving the evaluation and diagnosis, and providing the structure for the intervention.

The requisite knowledge for development of a systematic observation strategy (see the next section) includes how to observe, based on the overall movement or its phases (see Appendix 2.2), the best observation (or vantage) points, and how many observations are needed. Aids to the development of this strategy include the use of videography and rating scales. It is advisable for all movement analysts to practise observation even when using video, particularly if movements are fast and complex, as in notating games. Furthermore, the analyst should develop pre-pilot and pilot protocols to ensure all problems are overcome before the 'big day'. The maxim 'pilot, pilot, then pilot some more' is well founded.

Knowledge of effective instruction, feedback and intervention provide the appropriate information to translate critical features into intervention cues, couched in behavioural terms, which are appropriate to one's 'clients'. These should not be verbose – no more than six words – and figurative not literal. Remember that analogies must be meaningful; advising that the backhand clear in badminton is like 'swiping a fly off the ceiling with a towel' has no meaning to someone who has never performed such an action or seen another person do it. The cues to be devised can be verbal, visual, aural or kinaesthetic, and may differ for various phases of a movement; for example a javelin coach may see value in attending to the aural cues of footfall during the run-up, but would switch to other cues for the delivery phase. The movement analyst needs, therefore, to derive relevant cues for each movement phase, and should attend to: the cue structure (what the action is); its content (what does the action – the doers); and cue qualification (how to gauge success). Special conditions may be added if more information is needed. Examples include: rotate (action) the hip and trunk (the doers); swing (action) the arm (doer) forwards (qualification).

OBSERVATION STAGE – OBSERVING RELIABLY

As should be evident from Box 2.2, this stage primarily involves implementing the observation strategy devised in the preparation stage, and videographing the performers involved in the study. I use the term 'videographing' (or video recording) the performers advisedly; first, considerable skill is needed to observe reliably fast movements in sport by eye alone (see Table 2.1) and, secondly, good digital video cameras are now readily available and not expensive.

We need to record sports movements as they are fast and the human eye cannot resolve movements that occur in less than 0.25 s. Two important benefits of videography are that the performers can observe their own movements in slow motion and frame by frame, and that it makes qualitative analysis much easier. However, there are some potential drawbacks. Performers might be aware of the cameras and, consciously or subconsciously, change movement patterns (the Hawthorne effect). Also, there are ethical considerations about video recording, particularly with minors and the intellectually disadvantaged. Our systematic observation strategy should have addressed both what to focus on and how to record, and observe, the movements of interest. Clearly, we should focus on the critical features of the movement identified in stage 1, but we need to prioritise these. Secondly, we need to decide on the environment in which to videograph, the best camera locations within that environment and how many trials of the movement to record for analysis.

Prioritising critical features can vary with the skill of the performer, the activity being analysed, and whether a movement-phase approach (Appendix 2.2) is used, as in the long jump example later in this chapter (pages 62–71). Our prioritising strategy might, for example, put the critical features in descending order of importance for the performance outcome; or work from the general to the specific, for example from the whole skill to the role of the trunk and the limbs; or focus on balance, in skills in gymnastics.

The other main issues in videography for qualitative analysis are:

- Choice of camera shutter speed.
- Where to conduct the study.
- Choice of camera locations (sometimes, particularly in North America, called vantage points) and whether the cameras are to be stationary (usually mounted on tripods) or moved to follow the analysed movements.
- How many trials to record, when relevant.
- Use of additional lighting, which must be adequate for the shutter speed and frame rate. The latter is normally fixed for 'domestic-quality' video cameras, at 50 fields per second in Europe and 60 in North America, or 25 frames per second in Europe and 30 in North America (the unit hertz, Hz, is normally used for events per second).
- Who and what to observe.
- The background should be plain and uncluttered to help objective observation, but this is not always feasible, particularly when videographing in competitions.
- Participant preparation – briefing, clothing, habituation, debriefing.
- Size of the performer on the image – the bigger the better, but this might require zooming the camera lens (assuming that your camera has a zoom lens) while also panning and tilting the camera during filming.
- Checks for reliability (within, or intra-, observer) and objectivity (among, or inter-, observer) in any study.

So, let us now look at these points in a little more detail. The shutter speed is the time that the camera shutter stays open for each 'picture' that the camera records. If too slow, the picture will blur; if too fast for the lighting conditions, the picture may be too dark. A guide to the slowest satisfactory shutter speeds is given in Table 2.1. Not all digital video cameras for the domestic market will have the fastest of these shutter speeds, and do regard with suspicion the 'sports' option that cheaper cameras tend to use rather than a range of shutter speeds. We should also note that field rates, also known as sampling rates, of 50 or 60 Hz are far from ideal for the fastest activities in

Table 2.1 Examples of slowest satisfactory shutter speeds for various activities

ACTIVITY	SHUTTER SPEED (s)
Walking	1/50
Bowling (lawn or tenpin)	1/50
Basketball	1/100
Vertical jump	1/100
Jogging	1/100 to 1/200
Sprinting	1/200 to 1/500
Baseball pitching	1/500 to 1/1000
Baseball hitting	1/500 to 1/1000
Soccer kicking	1/500 to 1/1000
Tennis	1/500 to 1/1000
Golf	1/1000 or faster

Table 2.1. However, most movement analysts do not have routine access to high-speed video cameras with sampling rates up to thousands of pictures per second. If your needs analysis shows a clear requirement for such cameras, then this should be factored into the project costing in the preparation stage.

When deciding where to conduct the study, we have to balance an environment in which we have control over extraneous factors, such as lighting and background, and one that is similar to that in which the movement is normally performed; the latter ensures ecological validity. Normally, the latter dominates, but the decision may be affected by the skill of the performers, whether the activities being recorded are open or closed skills, and videographic issues. When selecting camera vantage points, the movement analyst has to address from where he or she would want to view these activities for qualitative analysis, with how many cameras, and whether the cameras need to be stationary.

The decision of how many trials, or performances, to record is very important for the reliability of qualitative analysis. However, that decision is not always made by the movement analyst. For example, if you were recording from a game, say of football, for notational analysis, you only have control over how many games you will record. If recording for technique analysis in competition, the number of recordable trials is probably fixed, for example, at six throws in the finals of a discus competition, the heats plus the finals of swimming events, and as many attempts as the jumper needs in the high jump until three failures. If recording out of competition, we need to decide how many observations we need; generally, within reason, the more the better. Because of movement variability, there is no such thing as a representative trial even for stereotyped closed skills. The more trials we record, the more likely are our results to be valid. Various rules of thumb have proposed between five and twenty trials as a minimum requirement; ten, if you can record that many, is often highly satisfactory.

Finally, we need to ensure that, in this stage, we attend to issues that affect our ability to assess, and improve, intra- and inter-analyst reliability. Reliability is consistency in ratings by one analyst, so we need to be able to check this over several days. Objectivity is consistency in ratings across several analysts, so we need enough analysts to be able to check this; clearly, this will be affected by how well trained the analysts are. Our assessments of objectivity and reliability can be improved by identifying critical features and how, and in which order, they will be evaluated; developing specific rating scales; analyst training and practice; and increasing the number of analysts or trials.

EVALUATION AND DIAGNOSIS STAGE – ANALYSING WHAT'S RIGHT AND WRONG IN A MOVEMENT

The hard work for this stage should already have been completed during the preparation stage – the identification of the critical features of the movement. The observation stage should then have allowed you to collect the video footage you need to evaluate these critical features in the performances that you have recorded. This stage also prepares us for the intervention stage. Often, in the evaluation and diagnosis stage – probably the most difficult of the four-stage process – you will start by describing the movement and progress to analysing it; trying to analyse a movement before you have thoroughly and scientifically described it can be fraught with difficulties. In this context, it should be noted that the work we do in this stage can do more harm than good; that is, we could reduce performance or increase injury risk, particularly if we have not identified and prioritised the correct critical features. This overall stage could be called the analysis stage; however there are two separate aspects to this stage (although they often overlap):

- To evaluate strengths and weaknesses of performance (what the symptoms are).
- To diagnose what weaknesses to tackle and how (diagnose symptoms and prepare to treat the condition).

Evaluation of performance

To evaluate performance we effectively need to compare the observed performances with some model of good form. However, as there is no general optimal performance model, we need a model that is appropriate for the performers being evaluated – the model needs 'individual specificity'. This clearly requires prior identification of critical features in the preparation stage. Furthermore, a ranking of the 'correctness' of the identified critical features on some scale or within some band of correctness can be very helpful; for example, 'joint range of motion: inadequate; within good range; excessive'; or 'excellent 5 . . . . OK 3 . . . . poor 1'. As well as needing a 'model' that is individual-specific, other difficulties arise in the evaluation of performance. The first of these relates to within-performer movement and performance variability; as we noted in the observation stage above, this can only be accounted for by recording multiple trials. Identifying the source of movement errors can also be problematic as they can arise from: body position or movement timing (biomechanical); conditioning (physiological); the performer evaluating environmental cues (perceptual-motor); or motivational factors (psychological). These factors support the need for movement analysts to be able to draw on a range of disciplinary skills and knowledge. In the real world of sport, movement analysts are usually most effective when they work as part of a multidisciplinary team of experts. Analysis bias, reliability and objectivity also present problems. Bias can be reduced by the use of 'correctness' criteria. Assessing reliability and objectivity requires multiple trials or analysts respectively; the latter is often a luxury, the former is vital.

Diagnosis of movement errors

Perhaps the major issues in the evaluation and diagnosis stage relate to the lack of a consistent rationale for diagnosing movement errors: our 'critical features' approach is best, providing that we can identify and prioritise the correct critical features. As only one intervention at any time is best, in the intervention stage, we need to focus on one correction at a time. This raises the question of how we diagnose to prioritise intervention. Five approaches are used, depending on the activity and circumstances.

The first of these focuses on 'what came before', in other words the relationship to previous actions, as in a stroke sequence in tennis. The second, somewhat related to the first, looks at the correct sequence through the phases of the movement (see Appendix 2.2). These two approaches are conceptually attractive, as problems usually arise before they are spotted. For example, in our long jump model below, landing problems are often due to poor generation of rotation on the take-off board or control of it in the air. Some problems arise in implementing these approaches for complex multi-segmental sports movements. We need to be aware that body segments interact, such that muscles affect even joints they do not cross. For example, it is normal to record a lack of triceps brachii activity in the action phase of baseball pitching, even though this muscle group is the main extensor of the elbow. In throwing and kicking, it is not entirely clear if a proximal segment speeds up a distal one or a distal one slows down a proximal one.

The third, and perhaps the most obvious, approach seeks to prioritise the critical features that maximise performance improvement. To use the long jump model again, if a long jumper is not jumping far, speed is overwhelmingly the most important factor; so what critical feature would we prioritise to maximise performance? Run-up speed obviously. However in many cases, it is not at all easy to know what will maximise improvement; furthermore, we often need to balance short-term and long-term considerations. In terms of successful outcomes, a fourth, and very attractive, approach is to make the easiest corrections first, in order of difficulty. This is impeccably logical from a motor skills viewpoint if movement errors seem unrelated and cannot be ranked. However there is little, if any, clear support for its efficacy in improving performance.

Finally, for activities in which balance is crucial, such as gymnastics and weightlifting, we might prioritise from the base of support upwards. But would this approach work, for example, in target shooting? From much experience, I would normally recommend to students of movement analysis the 'correct sequence' or 'what came before' approach to prioritising changes.

INTERVENTION STAGE – PROVIDING APPROPRIATE FEEDBACK

We come now to the final (intervention) stage of our four-stage process of qualitative analysis. Before getting this far, the movement analyst must have conducted a means analysis with the performer and their coach or therapist, identified the critical features relevant to the question to be answered, and prepared how feedback will be provided, including, for example, key words. Secondly, the movement analyst will have obtained relevant video footage and any other movement patterns (see Chapter 3 for discussion of the latter of these). Finally, the movement analyst will have analysed the video and movement patterns and prioritised the critical features to be addressed.

The focus in this final stage is on feedback of information to address the requirements of the needs analysis. If the previous three stages have been done badly, nothing in the intervention stage will sort matters out. On the other hand, provision of inappropriate feedback can jeopardise even good work done in the previous stages.

Several key points relate to providing information feedback. The information fed back should augment that available to the performer from his or her senses; such information is referred to, particularly in motor learning, as augmented feedback. The success of any intervention strategy to improve performance hinges on the way information is provided – fed back – to practitioners. The movement analyst must address what information is communicated, how this is done and when; this should have been partly done in the preparation stage. Practitioners may not always be receptive to feedback, particularly if it is not obviously relevant to the problems that a needs analysis should have identified; this difficulty often arises when no needs analysis has been carried out.

Some fundamental points should be borne in mind about providing feedback. First, we need accurate and reliable information to be fed back. Secondly, the information should provide something that is not directly observable by coaches or other practitioners. Thirdly, what is fed back should relate clearly to differences between good and poor performance. Fourthly, feedback should involve the right information at the right time and in an easily absorbed format. Lastly, the rapidity with which feedback is provided, its presentation and interpretation are all important.

It is worth noting, in this context, that the implicit assumption that feedback is inherently good is not totally supported. Further warning points are, first, that much information relating to movement technique is available, with little clarity about what should be fed back or how. Secondly, providing more information may cause confusion, particularly if unrelated to the problem identified. Thirdly, calls for the provision of immediate feedback, directly after the performance, don't address several very important points. The first of these is that the rapidity of feedback provision may depend on its role and may be different for a technique change to improve performance than for feedback of simple notational data. Next, feedback provision needs to address relevant motor learning research, particularly that of the ecological school. Too few research studies have addressed these issues in sport.

As an example of how movement analysts and coaches have got feedback wrong, let us look at how views of the generation of front crawl propulsive force have changed over the years – the wrong view often led to swimmers being given the wrong feedback. In the early 1970s, the predominant view was that the hand behaved as a paddle, pushing the water back – this led to coaches instructing swimmers to pull the hand back below the body in a straight line. By the mid-1970s to the 1980s, swimmers' hands had been shown, through cinematography, to make an S-shaped pattern through the water, in an outward–inward–outward sculling pattern. The hand was now envisaged as a hydrofoil, using lift and drag forces (see Chapter 5) to generate propulsion. Coaches now emphasised to their swimmers the need to develop a 'feel' for the water and to use sideways sculling movements. By the mid-1990s, these observed sideways movements had been shown to be due to body roll; the pattern of the hand relative to the swimmer's frame of reference – his or her body – consisted of an outward–inward scull, confounding the previous view, which had adopted a frame of reference fixed in the camera or swimming pool.

If provision of feedback involving knowledge of performance is good, then movement analysts need 'models' against which to realistically assess current performance. Alas, there is no general agreement about how we establish a model performance, technique, or movement pattern. Such models clearly need to fulfil the following functions: comparing and improving techniques, developing technique training, and aiding communication. As there is no such thing as a general optimal performance model, we have already noted that any such 'model' must be individual-specific.

Several further issues about feedback revolve around the questions: When is best? What is best? How is best? Immediate feedback is not necessarily best, as demonstrated by some of the literature on motor learning. For discrete laboratory tasks, summary feedback (of results) after several trials has been found to be better than immediate feedback for the retention stage of skill learning, which is the important stage – we want our performers to perform better the next day or week, rather than straight after feedback has been provided. However, it is still not clear if this applies to sport skills, which are far more complex than discrete laboratory tasks. Some evidence is contradictory; for example no difference was found for learning modifications to pedalling technique by inexperienced cyclists when they were provided with feedback from force pedals. However, do studies that relate to early skill learning also generalise to skilled performers? The jury still seems to be out on this one.

We might also ask whether the picture changes if we accept the views of ecological motor control. The constraints-led approach has supported the contention that an external focus of attention on the movement effects is better than an internal focus on the movement dynamics. The emphasis on task outcomes allows learners to search for task solutions and does not interfere with the self-organisational processes of movement dynamics, in contrast to the use of a movement-focused emphasis. This view has been supported by research in slalom skiing, tennis and ball kicking in American football. These provide evidence that 'less is more' – better performance results from less frequent augmented feedback. However, it is still unclear whether these results generalise to all stages of skill acquisition, particularly for highly skilled performers. The finding that a focus on movement dynamics is worse than one that places more emphasis on outcomes shows that movement analysts must be careful not to lose a focus on the movement outcome and must tailor technique feedback accordingly – this supports the qualitative approach to movement analysis.

Let us now address issues relating to the question, what is best? It is now technologically easy to feed back immediately 'kinetic' information, such as the forces on the pedals in cycling or the forces on the oars in rowing, or to use virtual reality, for example, to simulate bob-sled dynamics. The assumption is often made that immediate feedback of such information must improve performance. We have already noted that no real evidence supports this assumption as a retentive element in skill learning. Providing kinetic information seems to conflict with ecological motor learning research touched on above. Also, kinetic information might not relate to the performer's immediate problem.

Finally, how is best? Well, as simply as possible, using qualitative information that is easy to assimilate, preferably provided graphically and with appropriate cues, perhaps supported with some semi-quantitative data, such as phase durations, ranges of movement, or correctness scores. We should avoid complex quantitative data – if we do provide quantitative feedback, it should be graphical rather than numerical. Any feedback needs to be concise. It is very beneficial to compare good and bad performances, as in qualitative video analysis packages such as Dartfish™ (Dartfish, Fribourg, Switzerland; http://www.dartfish.com) and siliconCOACH™ (siliconCOACH, Dunedin, New Zealand; http://www.siliconcoach.com). Finally, it is advantageous if feedback can be provided in a take-away format, using DVDs or video tape, as can be done with, for example, siliconCOACH™.

The final task of this stage is to review the success of the project. This should be done independently, at least in the first instance, by the movement analyst and the 'clients'. The former should evaluate the whole four-stage process in the context of the needs analysis carried out in the preparation stage. The analyst and practitioners should then come together to discuss how improvements could be made in future studies. It is worth noting in this context that very few studies have addressed the efficacy of interventions by movement analysts. Although this is very difficult to establish, far too little attention is paid to addressing this important issue; after all, what is the point of the intervention stage if it doesn't have some provable benefit to performers?

IDENTIFYING CRITICAL FEATURES OF A MOVEMENT

Much of our work as movement analysts involves the study and evaluation of how sports skills are performed. To analyse the observed movement 'technique', we need to identify 'critical features' of the movement. These features should be crucial to improving performance of a certain skill or reducing the injury risk in performing that skill – sometimes both. For a qualitative biomechanical analyst, this means being able to observe those features of the movement; for the quantitative analyst, this requires measuring those features and often, further mathematical analysis (Chapters 4 to 6). Identification of these critical features is probably the most important task facing a qualitative or quantitative analyst, and we will look at several approaches to this task in this section. None is foolproof but all are infinitely better at identifying these crucial elements of a skill than an unstructured approach. Sometimes it can be helpful to define a 'scale of correctness' for critical features, for example poor = 1 to perfect = 5, or a 'range of correctness', such as 'wrist above elbow but below shoulder'.

The 'ideal performance' or 'elite athlete template' approach

This involves devising a set of critical features identified from an 'ideal' (sometimes called a 'model') performance, often that of an elite performer, hence the alternative name. This approach has nothing to recommend it except, for a lazy analyst, its minimal need for creative thought. It assumes that the ideal or elite performance is applicable to the person or persons for whom the analyst is performing his or her analysis. There is now wide agreement among movement analysts that there is no universal 'optimal performance model' for any sports movement pattern. Each performer brings a unique set of organismic constraints to a movement task; these determine which movements, out of the many possible solutions for the task under those constraints, are best for him or her.

Movement principles approach

As we saw in a previous section, different authors propose different principles under-lying coordinated movements in sport. This does not provide a convincing backdrop for identifying critical features of any movement by reading down a list of such principles. Categorising principles as general, partially general, or specific (as in Box 2.3), while it does conform to the constraints-led approach (see below), does not, in my recent experience, necessarily provide the answer either.

Nevertheless, the 'list of principles' approach is commonly used and often works well when analysing low-skill individuals. However, it is very susceptible to blind alleys as relationships between critical features are not apparent, in stark contrast to the deterministic modelling approach. I would caution against a mechanistic application of the movement principles approach. I would advise instead awareness of the important movement principles that need to be used in devising a deterministic model of a given

BOX 2.3 SUMMARY OF UNIVERSAL AND PARTIALLY GENERAL MOVEMENT PRINCIPLES (see Appendix 2.2 for details of those in the first two categories)

Universal principles

These should apply to all sports tasks:

- Use of the stretch–shortening cycle of muscle contraction.
- Minimisation of energy used to perform the task.
- Control of redundant degrees of freedom in the segmental chain.

Partially general principles

These apply to groups of sports tasks, such as those dominated by speed generation:

- Sequential action of muscles.
- Minimisation of inertia (increasing acceleration of movement).
- Impulse generation or absorption.
- Maximising the acceleration path.
- Stability.

Specific principles

These apply to the specific sports task under consideration and are derived and used for the long jump on pages 62–71 (see also Study task 5).

BOX 2.4 LEAST USEFUL MOVEMENT PRINCIPLES (IN MY EXPERIENCE)

- Summation of joint torques – cannot be observed and rarely measured, or estimated, accurately.
- Continuity of joint torques – as above.
- Equilibrium or stability – grossly oversimplified principle for fast movements.
- Nature of segments – I have never been sure what this means.
- Compactness – another grossly oversimplified principle for fast movements.
- Spin – say no more!

sports movement, or selecting the principles that conform to the constraints on the movement, by applying Box 2.3 for example. The principles used should be specific to the sport, the performer and the constraints on the movement. With a very sharp warning that any set of movement principles is neither a list to be run through in all circumstances nor a 'cookbook', the principles in Box 2.3 are those that I have found most useful in devising deterministic models for sporting activities in which I have been involved as an analyst. These activities include athletic throwing and jumping events, cricket, basketball, hockey and rugby. The lack of any 'acrobatic' sports in the last sentence probably colours the inclusion or otherwise of some movement principles in Box 2.3 (and 2.4). I also include in Box 2.4, somewhat tongue-in-cheek, some so-called movement principles that I have not found very useful in my work.

Deterministic modelling

Principles of deterministic modelling

Deterministic models of sports activities, also known as hierarchical models as they descend a hierarchical pyramid, can be developed using a structure chart, for example that in Microsoft PowerPoint (Microsoft Visio, probably less familiar to most readers, is far better for such modelling). The first principle of hierarchical modelling is to identify the 'performance criterion', the outcome measure of the sporting activity. This is often, in track and field athletics for example, to go faster, higher or further. In such cases, we have a clear and objective performance criterion, such as race time, which we seek to minimise, or distance jumped or thrown, which we seek to maximise. Splitting a movement into phases, as I have done below for the long jump (see also Appendix 2.2), can not only aid the establishment of a deterministic model but also help the identification of objective performance criteria for each phase.

In sports involving a subjective judgement, such as gymnastics, identifying a performance criterion may be far more difficult. The score awarded by the judges will not depend upon a single performance factor, whether objective or subjective, but on guidelines established by the sport. The constituent parts of such performances may be analysable by deterministic models, for example the performance of a twisting somersault or the pre-flight phase of a vault. On the other hand, they may be better approached using the judging guidelines of that sport, which are based mainly on technical elements of the skills involved; this approach will be discussed briefly on pages 71–2. Some sports, such as ski jumping, combine objective (distance) and subjective (style) criteria – the former lends itself to being modelled deterministically, unlike the latter.

The next stage is to subdivide the performance criterion, where possible, as in the example below. Then comes the crucial stage of identifying critical features, also known, particularly when the model is developed for quantitative rather than qualitative analysis, as performance variables or parameters. Once this is done, and the model has been developed to the necessary stage – which should arrive at observable features in a qualitative analysis or measurable ones for a quantitative analysis – it needs to be evaluated and its limitations noted. Generally, the critical features highlighted in the model will be biomechanical features or variables such as joint angles, or body segment parameters such as a skater's moment of inertia. Generally, it is advisable not to use ambiguities, such as 'timing' or 'flexibility' or even, perhaps, 'coordination'. If critical, these should be identified more precisely, such as specifying why hamstring flexibility is important because it improves the joint range of movement, or which joint movements need to be coordinated and how.

Model entry at one 'level' should be completely defined by those associated with it at the next level down, for example those at the top level 1 – this would be the performance criterion – by those at level 2. This association should either be a division – as in the example below (Figure 2.3) for the long jump distance, or a biomechanical relationship. The latter, of course, will require the analyst to be aware of movement principles (Appendix 2.1). The difference between this approach and using a list of principles, as above, is that the principle flows from the model rather than being slavishly adopted from a list of these things. As well, this modelling approach is easily adapted to alternative ways of identifying critical features, for example through a constraints-led approach (pages 71–2). An advantage of hierarchical models over lists of principles is that they help the movement analyst to spot 'blind alleys', as again illustrated in the following example.

Example: Hierarchical model for qualitative analysis of the long jump

The first step is to define the performance criterion, which is very simple for this task, being the distance jumped. We will ignore here compliance with the rules of the event – which are task constraints in their own right, assuming the jumper analysed is able to conform to the rules. The next step – level two of our model, as in Figure 2.3 – is to ask if we can divide the distance jumped into other distances, which might relate to the phases of the movement (see below and Appendix 2.2). Here, it can be subdivided into the take-off distance, the flight distance and the landing distance (these are explained in Figure 2.4).

We have now completed level 2 of our long jump model and need to prioritise further development according to which of the three sub-distances is most important. Clearly, from Figure 2.4, the flight distance is by far the most important and is the distance the jumper's centre of mass (see Chapter 5) travels during the airborne, or flight, phase. This distance can be specified by a biomechanical relationship as it fits the model of projectile motion (see Chapter 4). The determining biomechanical parameters of take-off are the take-off speed, take-off angle and take-off height, as in Figure 2.5, which is level 3 of our deterministic model for the flight distance. This distance is also seen to be affected by the aerodynamics of the jumper – the air

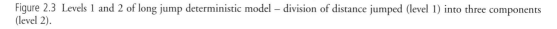

Figure 2.3 Levels 1 and 2 of long jump deterministic model – division of distance jumped (level 1) into three components (level 2).

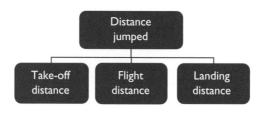

Figure 2.4 Explanation of division of distance jumped into three components: TOD = take-off distance; LD = landing distance; circle denotes position of jumper's centre of mass.

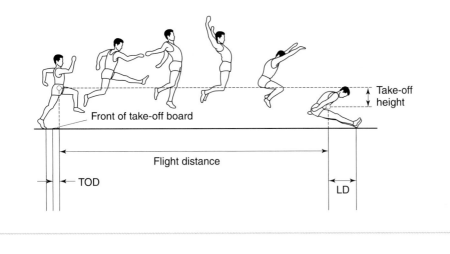

Figure 2.5 Level 3 of long jump model – factors affecting flight distance.

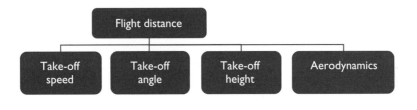

resistance. A quantitative analyst would probably ignore air resistance, having no easy way of measuring it. The qualitative analyst must be careful not to be led down a 'blind alley' here; air resistance could be reduced by adopting a tuck or piked position, which would be more 'streamlined' (see Chapter 5) than the extended body position of the jumper in Figure 2.4. However, although many novice long jumpers tend to pike or tuck, such a position is detrimental to overall performance, as it encourages forward rotation of the jumper's body, which adversely affects the landing distance (see above).

Again, we now need to prioritise the development of level 4 of the model. Which of the take-off parameters is most important? Most readers will know that the take-off speed is by far the most important, as the distance jumped is roughly proportional to the square of the take-off speed, and that the other two take-off parameters are far less influential (see Chapter 4 for confirmation of this). Level 4 for the take-off speed is shown in Figure 2.6(a), where the take-off speed has been divided into the run-up speed and the speed added – or lost – on the take-off board. The first of these gives us our first critical feature – the jumper must have a fast run-up; we need to develop an 'eye' for this as qualitative analysts whereas a quantitative analyst would need to devise the best way of measuring this speed. The second component – speed added on the board – is of little use, being unobservable (although measurable), so we need to replace it by a biomechanical relationship.

The one relationship that should spring to mind most easily – because it is the simplest – is the impulse–momentum relationship outlined in Appendix 2.1 (see pages 77–8) and explained in more detail in Chapter 5. This is represented graphically in Figure 2.6(b), in which the speed added on the board has been replaced by the take-off impulse and the athlete's mass (at the same level of the model). Take-off impulse depends at the next level down on the mean force and the time on the take-off board. This figure may not seem much of an advance on Figure 2.6(a) – the jumper's mass is no problem but observing take-off impulse, time on board or mean force is down-right impossible although, interestingly, all of these are easily measurable if the take-off board is mounted on a force plate (see Chapter 5). In a way, this impossibility is fortunate for the qualitative analyst, as this 'branch' of the model is a blind alley, down which the quantitative analyst could easily wander. The reason is that this branch of the model implies that we need to maximise both the mean force and the time on the board to improve performance. Given the fast run-up of good long jumpers, trying to extend the board contact time (about 15 ms) would be difficult at best; if feasible, it would have a deleterious effect on the mean force. Fortunately we have escaped from this blind alley simply by being unable to observe these features; that is not always the case.

A better approach altogether is to use the work–energy relationship outlined in Appendix 2.1 (see page 78). This is shown in the new version of level 4 (and level 5) of

Figure 2.6 Level 4 of long jump model – factors affecting take-off speed: (a) initial model; (b) 'blind alley'.

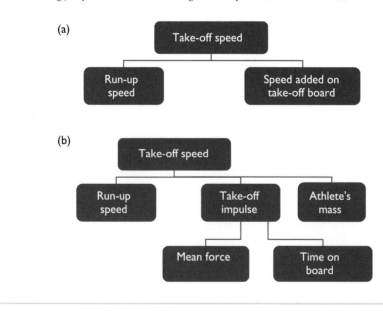

Figure 2.7 Level 4 of long jump model – factors affecting take-off speed – avoiding the blind alley.

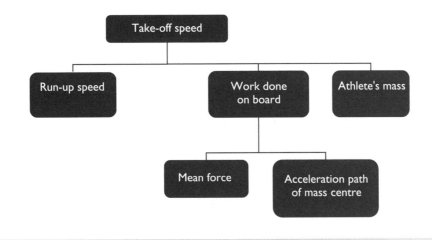

our model, in Figure 2.7. Although mean force and acceleration path of the jumper's centre of mass won't yet appear to be observable, we do have the right relationship, as the jumper correctly needs to try to maximise both of these factors.

Figure 2.8 Take-off velocity components.

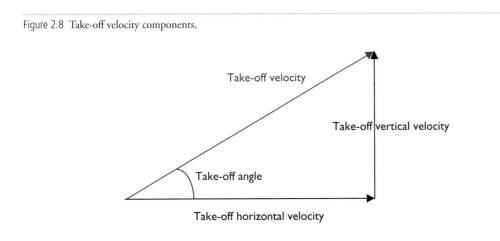

Now all we have to do is develop observable features that capture these bio-mechanical entities. We will take a step aside for the moment, as the observable features we need flow more naturally from considering the vertical and horizontal components (in black) of the take-off velocity (see Figure 2.8), rather than the take-off speed, which is simply the magnitude of the overall take-off velocity in Figure 2.8; the take-off angle specifies the direction of the take-off velocity. Hence, the take-off velocity can be specified by either the take-off speed and angle, as in Figure 2.5, or the horizontal and vertical take-off velocities, as in Figure 2.9. The latter approach is much better here, indeed it is crucial, because what happens on the take-off board differs significantly for the horizontal and vertical velocities. On the board, a long jumper needs to generate vertical velocity while minimising the loss of horizontal velocity from the run-up. For the horizontal component, there tends to be an initial 'braking' force followed by an accelerating force. We also need these two velocity components to develop our model of take-off angle, the tangent of which, from Figure 2.8, is simply the vertical velocity at take-off divided by the horizontal velocity at take-off. However, with the modified approach, we do not need to develop a separate branch of our long jump model for take-off angle, as we have now effectively replaced level 3 of the model for flight distance, in Figure 2.5, by Figure 2.9.

Figure 2.9 Revised long jump model for flight distance.

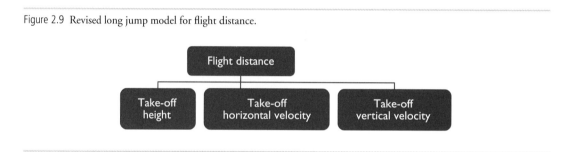

We now, therefore, replace the branch for take-off speed (level 3) in our long jump model (Figure 2.6) by separate branches for the take-off horizontal and vertical velocities. We then develop levels 4 down of our model for each of these branches by using relationships for the two components of the take-off velocity that are equivalent to the ones we developed above for take-off speed; this gives us Figures 2.10 and 2.11.

Figure 2.10 Factors affecting take-off horizontal velocity.

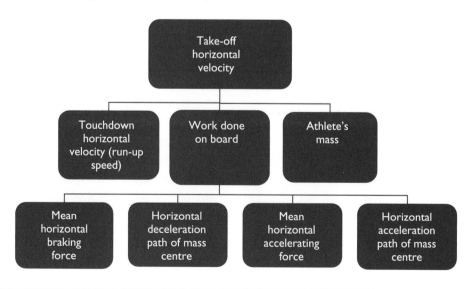

Figure 2.11 Factors affecting take-off vertical velocity.

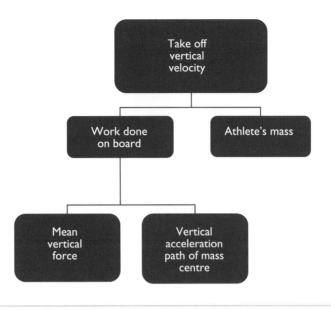

At this stage, to remove what are now superfluous details so as to clarify the model, we delete the 'work done on board' boxes of our long jump model, as these are completely specified by the level below, and omit the 'athlete's mass' box, as this is rather trivial. As the take-off velocity is, as in Figure 2.8, made up of the take-off horizontal and vertical velocities, we can combine these simplified figures into the final branch of our long jump model for take-off velocity (vertical and horizontal velocity and also take-off angle) of Figure 2.12.

We now need to consider what observable, critical features of the movement contribute to the lowest level of each branch of this model. We depart slightly from slavish adherence to the principles of hierarchical modelling at this stage, but we must ensure that we can propose biomechanical, physiological or other scientific principles to justify these lower levels of the model.

Figure 2.12 Final model for take-off vertical and horizontal velocities.

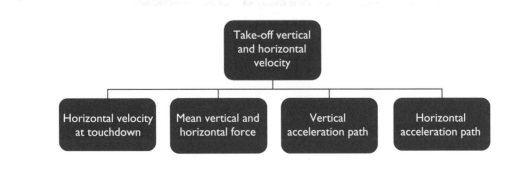

Figure 2.13 Identifying critical features that maximise force generation and vertical and horizontal acceleration paths. (Notes: [1]also covers take-off height; [2]blind alley).

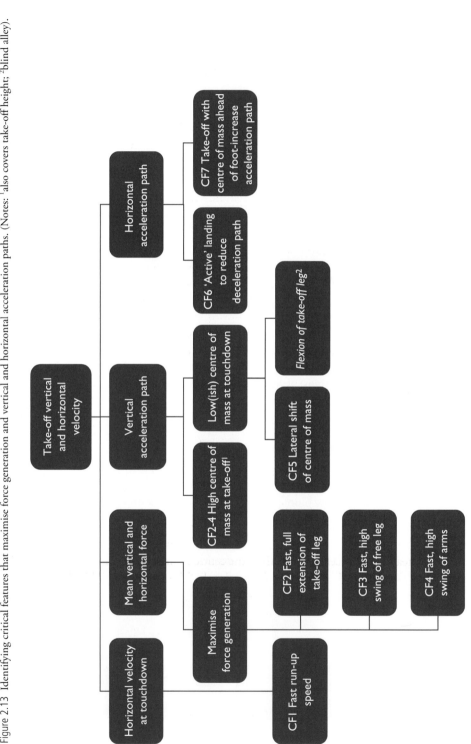

Let us start at the left branch of the model in Figure 2.12 and work to the right. We have already identified 'fast run-up speed' as a critical feature (CF1) for this event. We have already noted that maximising the mean force and the acceleration path are desirable to maximise take-off speed (but see below), so we now need only to translate these terms into things we can observe. The mean forces are maximised by the jumper maximising force generation (Figure 2.13). This can be done in three ways: directly, by a fast and full extension of the take-off leg (CF2) increasing the force on the take-off board and, indirectly, by a fast, high and coordinated swing of the free leg (CF3) and the two arms (CF4). If the indirect contributions are not clear, refer to Figures 1.21 and 1.22 (pages 24–5), in which the normal and model swings of the arms in the standing vertical jump both increased jump height.

Moving on to maximising the vertical acceleration path, we first note that this is expressed as the difference between the heights of the athlete's centre of mass at take-off and touchdown. The jumper can achieve a high centre of mass at take-off by a combination of critical features 2 to 4 (CF2–4). A lowish centre of mass at touchdown might suggest a pronounced flexing of the knee at touchdown. Although knee flexion will occur to some extent and this will reduce impact forces and thereby injury risk, it would be a mistake for the jumper to try to increase this flexion – it would lower the centre of mass height at touchdown but have far more important and deleterious effects on the take-off speed. A mechanism that good long jumpers tend to use to lower the centre of mass at touchdown is a lateral pelvic tilt towards the take-off leg. This is clearly evident from a front-on view and illustrates two important points for a successful qualitative movement analysis: know your sport or event inside out and never view a sporting activity just from the side, even when it seems two-dimensional.

Now let's consider the horizontal acceleration path. This is more tricky because, in the first part of board contact, until the centre of mass has passed forward of the support foot, the horizontal velocity will be decelerating, not accelerating. The last thing the jumper would want to do is to plant the take-off foot too far ahead of the centre of mass, which would increase the deceleration of the centre of mass – yet another blind alley. Instead, the jumper minimises this distance by seeking an 'active' landing – one in which the foot of the take-off leg would be moving backwards relative to the take-off board at touchdown – to reduce the horizontal deceleration (CF6). Then, once the centre of mass has passed over the take-off foot, the jumper needs to lengthen, within reason, the acceleration path of the centre of mass up to take-off, which can be done by taking off with the centre of mass ahead of the foot (CF7). This also serves to minimise any tendency for the take-off distance (Figure 2.4) to be negative.

Well, we have finished with take-off velocity, which covers both take-off speed and angle. Take-off height (level 3 of the model, see Figure 2.5) is mainly dealt with by critical feature 4 (CF4), which has now covered flight distance from level 2 of the long jump model – although we still need to look at the landing component of the take-off height, which we will do in the next paragraph – leaving us with the take-off and landing distances.

From Figure 2.4, it should be obvious that the take-off distance is the distance of the centre of mass in front of the take-off foot at take-off minus the distance of the take-off foot behind the front of the take-off board, from which the jump distance is measured. We have already considered how to increase the first of these distances (CF7, Figure 2.13), so we now note the need to plant the take-off foot as close to the front of the take-off board as possible (CF8, Figure 2.14). This has implications for the control of the run-up, so we might wish to amend our first critical feature (CF1) to 'fast and controlled run-up'. The landing distance (Figure 2.4) is the distance of the centre of mass behind the feet at landing minus the distance that the point of contact of the

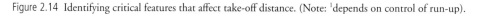

Figure 2.14 Identifying critical features that affect take-off distance. (Note: [1]depends on control of run-up).

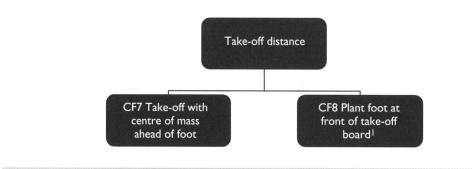

Figure 2.15 Identifying critical features that affect landing distance. (Note: [1]distance of centre of mass behind feet at landing minus distance of any other contact with sand behind feet after landing).

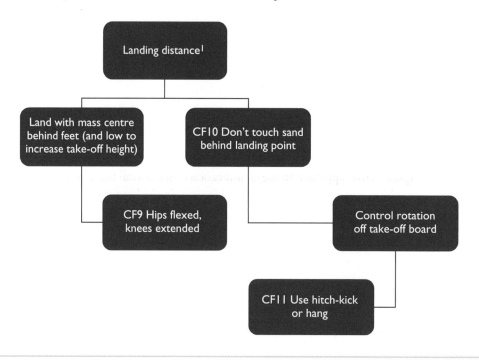

jumper's body closest to the take-off board is behind where the feet land. That leads to Figure 2.15, which contains our last three critical features. First, from the observation that the athlete should land with his or her centre of mass behind their feet and low to increase take-off height, we derive the need to land with 'hips flexed and knees extended' (CF9), which is evident from Figure 2.4. To reduce any loss of landing distance from touching the sand behind the landing point for the feet, we note simply 'don't touch the sand behind the landing point' (CF10). The final critical feature is less obvious and emphasises two important points: the need to be aware of the movement principles relevant to the activity analysed and to have a thorough knowledge of that activity. The forces acting on the jumper from the take-off board generate 'angular momentum' that tends to rotate the jumper forwards during flight. If uncontrolled, this would cause an early landing in the pit, which is why tucking or piking during flight are counterproductive. Instead, the jumper needs either to minimise forward rotation by adopting an extended 'hang' position, as in Figure 2.4 or, for longer jumpers, to transfer this angular momentum (see Chapter 5) from the trunk to the limbs using a 'hitch-kick' technique, leading to our last critical feature 'use hitch kick or hang' (CF11).

Well we've got there, although it may have seemed a long journey. It is worthwhile, because we finish up with confidence in our critical features from the rigour of the deterministic modelling process, which is impossible to achieve from the copying of an 'ideal' or 'model' performance, very difficult to achieve merely from a list of movement principles, and not always clear from other approaches. Also, as already noted, the process highlights blind alleys, helping us to avoid them, and provides a well-structured approach for identifying critical features.

Summary of the use of deterministic models in qualitative movement analysis

- Using diagrammatic deterministic models is, in many cases, the best approach to identifying critical features of a movement if we can formulate a clear performance criterion.
- It helps to overcome many pitfalls of qualitative analysis, such as a lack of a structured approach to identifying critical features even within an overall structured approach, and wandering down blind alleys.
- It can suffer from rigid formalism; however, the step from the initial 'qualitative or quantitative' approach to the identification of the critical features to be observed in a qualitative analysis allows greater freedom, as in the long jump example above.
- This greater freedom can be a problem as well, which can only be avoided by thorough knowledge of the activity being analysed and from an awareness of relevant movement principles and the constraints on the movement.

Other approaches to identifying critical features

In Chapter 1, we touched on the constraints-led approach to human movement analysis. In this relatively new approach, a particular sports movement is studied as a function of the environmental, task and organismic constraints on that movement. This approach recognises that each sports performer brings to a specific movement task, such as throwing a javelin, a set of organismic constraints unique to that person. These determine which movement patterns, from the many possible solutions to the task and environmental constraints, are best suited to that individual. Environmental constraints are largely related to rules, equipment and, unsurprisingly, the environment in which the activity occurs. Organismic constraints include anatomical and anthropometric factors and fitness. Task (or biomechanical) constraints include the forces and torques needed to perform the movement plus inertia, strength, speed and accuracy. This approach has not yet been developed sufficiently by sports biomechanists to be an alternative way of identifying critical features of a movement. However, awareness of the constraints on a particular movement can help the movement analyst to identify critical features with more confidence, in combination with deterministic modelling or, perhaps, the movement principles approach.

The long jump example on pages 62–71 will have shown you that deterministic modelling is not a 'quick fix' for identifying critical features of a sports movement for qualitative analysis, even when the objective performance criterion is easily identified. However, although it takes some time to complete, it is not too difficult once you have practised it well and if you are very familiar with the sports movement involved. The same is true for quantitative analysts using hierarchical modelling to identify the important performance variables to be measured. When the performance variable is subjective, as in all sports in which subjective judging determines the outcome score (e.g. gymnastics, diving and figure skating), an obvious alternative is to base the critical features on the judging guidelines for the particular sport. These should largely have been developed from movement principles applicable to movements within that sport, so this approach has much to recommend it, particularly for inexperienced movement analysts.

SUMMARY

In this chapter, we considered how qualitative biomechanical analysis of movement is part of a multidisciplinary approach to movement analysis. We looked at several structured approaches to qualitative analysis of movement, all of which have, at their core, the identification of critical features of the movement studied. We identified four stages in a structured approach to movement analysis, considered the main aspects of each stage and noted that the value of each stage depends on how well the previous stages have been implemented. We saw that the most crucial step in the whole approach is how to identify the critical features of a movement, and we looked at several ways of doing this, but found that none is foolproof. We worked through a detailed example of the best approach, using deterministic models, and considered the 'movement principles' approach and the role of phase analysis of movement.

STUDY TASKS

1 From the bullet points for each of the four stages of our structured approach to qualitative movement analysis in Box 2.2, explain briefly the main issues requiring attention for each 'first level' bullet point that is not cross-referenced to another chapter.
Hint: You may wish to refer to the details of each of the four stages on pages 48–58 if you are struggling with this task.

2 Identify the movement phases into which the long jump can be divided, specify the boundaries of each phase, and outline the main functions of each phase (between one and three functions for each phase). Which three of the critical features on Figures 2.13 to 2.15 would most probably have the greatest effect on performance?
Hint: You may wish to refer to the deterministic model of the long jump in pages 62–71 and to Appendix 2.1 before undertaking this task.

3 Show, diagrammatically, the two main rules to be used in moving down one level in deterministic modelling.
Hint: You may wish to reread the subsection on 'Principles of deterministic modelling' (pages 61–2) before undertaking this task.

4 For one of the following sports activities, a movement phase approach to each of which is outlined in Appendix 2.2 (pages 81–2), identify the most important phase from a performance perspective and then derive, and display diagrammatically, levels 1 to 3 of a deterministic model for that phase. The activities are stroking in swimming, the volleyball spike and the javelin throw.
Hint: Before undertaking this task, you should read Appendix 2.2 (pages 78–82); you may also wish to reread the subsection on 'Principles of deterministic modelling' and those parts of the subsection 'Hierarchical model for qualitative analysis of the long jump' (pages 62–3) that deal with levels 1 to 3 of that model. You may also find useful aids in Hay (1993; see Further Reading, page 76).

For the next four study tasks, you will need to choose a sports activity – other than the long jump – in which you are interested. This need not be the same activity for each of these tasks, but you will see more continuity – and you will be closer to the approach movement analysts use in the world of sport – if you do stick to the same movement throughout these four tasks. I strongly advise you to select an activity with an objective, rather than a subjective, performance criterion, and one that is covered on the Coaches' Information Services (CIS) website (http://coaches-info.com/) run by the International Society of Biomechanics in Sports. Also, specify the age, sex and performance standard of the performer you will consider in the study tasks below. Assume that a needs analysis with the performer's coach has highlighted your need to identify the main factors that contribute to success and their prioritisation for an intervention strategy.

5 Identify and list all the sources (including the CIS) that you could use to gather evidence-based information about that particular movement. Summarise, in not

more than 1000 words, the main features of the movement that you would use in a qualitative analysis of your chosen performer. From Appendix 2.1 (pages 76–8), identify the universal and partially general movement principles that are most, and those that are least, applicable to your chosen activity; include also two activity-specific movement principles.

Hint: You are advised to reread the long jump model above (pages 62–71), read carefully the sources of information you use, and pay careful attention to the points in Appendix 2.1 and Boxes 2.3 and 2.4.

6 Use a deterministic model to identify about six observable critical features for performance of your chosen activity. You will need to develop fully, and represent diagrammatically, levels 1 and 2 of the model, but you should not need to follow every branch down further levels (as we did in the long jump example); you should focus on developing the boxes in level 2 that most affect performance.

Hint: You are advised to reread the subsection on the long jump model above (pages 62–71) and, when developing your deterministic model, to pay careful attention to the relevant principles that you have identified as relating to your chosen activity in Study task 5.

7 Devise a systematic observation strategy for your chosen activity, including recording location, number of cameras, their positions, any auxiliary lighting, camera shutter speed, performer preparation, and the required number of trials. Set out an instruction sheet for conducting an initial pilot study. Outline, briefly, how you might ensure validity, reliability and objectivity while minimising observer bias.

Hint: Before undertaking this task, you may wish to reread the section 'Observation stage – observing reliably' (pages 51–4) and, if necessary, read Chapter 5 in Knudson and Morrison (2002; see Further Reading, page 76).

8 Decide which approach, from the five outlined in the section on 'Evaluation and diagnosis stage – analysing what's right and wrong in a movement' (pages 54–6) you would use to prioritise, for intervention, three from your set of critical features from Study task 6. Explain why these three were chosen, and their order, and outline briefly how you would provide performance-improving feedback for each of these three critical features, including verbal cues.

Hint: Before undertaking this task, you are advised to reread the section on 'Evaluation and diagnosis stage – analysing what's right and wrong in a movement' (pages 54–6). You may also wish to read Chapters 6 and 7 in Knudson and Morrison (2002; see Further Reading, page 76). You might also find that the feedback-oriented chapters (1 to 3) in Hughes and Franks (2004; see Further Reading, page 76) contain useful information].

You should also answer the multiple choice questions for Chapter 2 on the book's website.

▋ GLOSSARY OF IMPORTANT TERMS (compiled by Dr Melanie Bussey)

Degrees of freedom Used in movement analysis for the set of independent displacements that specify completely the displaced or deformed position of the body or system. Used more broadly in motor learning and control.

Deterministic model A model linking mechanical variables with the goal of the movement; most often used in qualitative analysis.

Ecological motor learning Holds that all movements and actions are influenced or constrained by the environment. Environmental information is necessary to shape or modify the characteristics of movement to achieve specific actions or tasks.

Efficacy The ability to produce a desired amount of a desired effect.

Inertia The reluctance of a body to move.

Kinaesthesis The ability to sense proprioceptively (from sensory receptors within the body) the movements of the limbs and body.

Projectile An object (or person) – that has been flung into the air. See also **projection angle**, **projection height** and **projection velocity**.

Projection angle (release angle, take-off angle) The angle at which a projectile is released. See also **projection height** and **projection velocity**.

Projection height (release height, take-off height) The difference between the height at which a projectile is released and the height at which it lands. See also **projection angle** and **projection velocity**.

Projection velocity (release velocity, take-off velocity) The velocity at which a projectile is released, may be broken into horizontal and vertical components. The magnitude of the projection velocity, with no indication of its direction, is the projection (release, take-off) speed. See also **projection angle** and **projection height**.

Range The horizontal distance a **projectile** travels.

Redundant In the context of motor control, this is the duplication of critical movements in a system to increase the versatility of the system – there is more than one way to locate a system or segment in a given position.

Sequential movement A movement that involves the sequential action of a chain of body segments, often leading to a high-speed motion of external objects through the production of a summed velocity at the end of the chain of segments.

Stretch–shortening cycle A common sequence of joint actions in which an eccentric (lengthening) muscle contraction, or pre-stretch, precedes a concentric (shortening) muscle contraction.

Torque The turning effect, or moment, of a force; the product of a force and the perpendicular distance from the line of action of the force to the axis of rotation.

Trajectory The flight path of a **projectile** determined by the horizontal and vertical acceleration of the projectile and its projection speed, angle and height.

FURTHER READING

Hay, J.C. (1993) *The Biomechanics of Sports Techniques*, Englewood Cliffs, NJ: Prentice Hall. Some of the sports chapters (Chapters 8 to 17) contain deterministic models of various sports activities, which should be of interest to you as further examples of this approach; some will also help with your study tasks.

Hughes, M.D. and Franks, I.M. (2004) *Notational Analysis of Sport*, London: Routledge. Although written mainly from a notational analysis viewpoint, Chapters 1 to 3 contain valuable information about performance-enhancing augmented feedback.

Knudson, D.V. and Morrison, C.S. (2002) *Qualitative Analysis of Human Movement*, Champaign, IL: Human Kinetics. The first edition of this book was for many years one of the few real gems in the Human Kinetics list of sports science texts; the second edition continues that tradition. However, the authors have not yet welcomed a wider interpretation of qualitative movement analysis and a crucial (perhaps the crucial) 'critical feature' of skilled human movement – coordination – receives only one page reference in the index. The structured approach to movement analysis outlined in this chapter is covered in far more detail in Part II of Knudson and Morrison, while Part I sets the scene nicely and Part III outlines applications of their approach, with many diagrammatic examples. Highly recommended and well written.

Kreighbaum, E. and Barthels, K.M. (1996) *Biomechanics: A Qualitative Approach for Studying Human Movement*, New York: Macmillan. I find the approach taken by these authors overly mechanics-based; such an approach has turned so many students off sports biomechanics over the years. However, Chapters 13 to 16 have much to recommend them.

APPENDIX 2.1 UNIVERSAL AND PARTIALLY GENERAL MOVEMENT (BIOMECHANICAL) PRINCIPLES

Universal principles – these should apply to all (or most) sports tasks

- Use of the stretch–shortening cycle of muscle contraction.
  Also referred to as the use of pre-stretch; in performing many sports activities, a body segment often moves initially in the opposite direction from the one intended. This initial countermovement is often necessary simply to allow the subsequent movement to occur. Other benefits arise from the increased acceleration path, initiation of the stretch reflex, storage of elastic energy, and stretching the muscle to optimal length for forceful contraction – relating to the muscle's length–tension curve. This principle appears to be universal for movements requiring force or speed or to minimise energy consumption.
- Minimisation of energy used to perform the task.
  Some evidence supports this as an adaptive mechanism in skill acquisition, for example the reduction in unnecessary movements during the learning of throwing skills. The many multi-joint muscles in the body support the importance of energy efficiency as an evolutionary principle. However, little evidence exists to support this as a universal principle for sports tasks involving speed or force generation.
- Control of redundant degrees of freedom in the segmental chain.

This is also known as the principle of minimum task complexity. The chain of body segments proceeds from the most proximal to the most distal segment. Coordination of that chain becomes more complex as the number of degrees of freedom – the possible axes of rotation plus directions of linear motion at each joint – increases. A simple segment chain from shoulder girdle to the fingers contains at least 17 degrees of freedom. Obviously many of these need to be 'controlled' to permit movement replication. For example, in a basketball set shot players may keep their elbow well into the body to reduce the redundant degrees of freedom. The forces need to be applied in the required direction of motion. This principle explains why skilled movements look so simple.

Partially general principles – these apply to groups of sports tasks, such as those dominated by speed generation

- Sequential action of muscles.
  This principle – also referred to as the summation of internal forces, serial organisation, or the transfer of angular momentum along the segment chain – is most important in activities requiring speed or force, such as discus throwing. It involves the recruitment of body segments into the movement at the correct time. Movements are generally initiated by the large muscle groups, which produce force to overcome the inertia of the whole body plus any sports implement. The sequence is continued by the faster muscles of the extremities, which not only have a larger range of movement and speed but also improved accuracy owing to the smaller number of muscle fibres innervated by each motor neuron. In correct sequencing, proximal segments move ahead of distal ones, which ensures that muscles are stretched to develop tension when they contract; the principle appears to break down for long axis rotations, such as medial–lateral rotation of the upper arm and pronation–supination of the forearm, which occur out of sequence in, for example, the tennis serve.
- Minimisation of inertia (increasing acceleration of movement).
  This is most important in endurance and speed activities. Movements at any joint should be initiated with the distal joints in a position that minimises the moment of inertia, to maximise rotational acceleration. For example, in the recovery phase of sprinting, the hip is flexed with the knee also flexed; this configuration has a far lower moment of inertia than an extended or semi-flexed knee. This principle relates to the generation and transfer of angular momentum (see Chapter 5), which are affected by changes in the moment of inertia.
- Impulse generation or absorption.
  This principle is mainly important in force and speed activities. It relates to the impulse–momentum relationship (see also Chapter 5): impulse = change of momentum = average force multiplied by the time the force acts. This shows that a large impulse is needed to produce a large change of momentum; this requires a large average force or a long time of action. In impulse generation, the former

must predominate because of the explosive short duration of many sports movements, such as a high jump take-off, which requires power – the rapid performance of work (see below). In absorbing momentum, as when catching a cricket ball, the time is increased by 'giving' with the ball to reduce the mean impact force, preventing bruising or fracture and increasing success.

- Maximising the acceleration path.
  This principle arises from the work–energy relationship, which shows that a large change in mechanical energy requires a large average force or the maximising of the distance over which we apply force. This is an important principle in events requiring speed and force, for example a shot-putter making full use of the width of the throwing circle.
- Stability
  A wide base of support is needed for stability; this applies not only to static activities but also to dynamic ones in which sudden changes in the momentum vector occur.

APPENDIX 2.2 OTHER EXAMPLES OF PHASE ANALYSIS OF SPORTS MOVEMENTS

As noted in this chapter, phase analysis involves the breaking down of movements into phases. The phases have biomechanically distinct roles in the overall movement and each phase has easily identified phase boundaries, often called 'key events'. The duration of the phases can be determined – to the accuracy of the time resolution of the recording system – for comparison between and within performers (hence the other name, temporal analysis). Any phase analysis should start with a statement of the reason for performing the analysis, which should preferably be done in the preparation stage but can be delayed until later if video is recorded. The movement analyst then needs to specify phases, identity their boundaries and determine their distinctive biomechanical features. The examples below, in addition to that of the long jump in this chapter, should clarify various aspects of phase analysis.

Phase analysis of ballistic movements

Ballistic movements, which include throwing, kicking and hitting skills, subdivide into three phases: preparation or backswing; action; recovery or follow-through. In the tennis serve, for example, the key events are:

- The start of the preparation phase – the initiation of ball toss.
- The end of the preparation phase, which is also the start of the action phase – this is somewhat arbitrary, but could be defined as the racket reaching its lowest point behind the server's back.
- The end of the action phase and start of recovery – impact of the racket and ball.

- The end of the recovery phase, which is also rather arbitrary and could be defined as the end of the serving arm's follow-through or the start of the player's movement into the next stroke.

Here, the preparation phase has various important functions:

- It puts the body into an advantageous position for the action phase.
- It maximises the range of movement of the racket – it increases the acceleration path (see Appendix 2.1).
- It allows the larger segments to initiate the movement (see Appendix 2.1).
- It stretches the agonist muscles, increasing the output of the muscle spindle receptors, and allowing the storage of elastic energy (the stretch–shortening cycle, see Appendix 2.1).
- It increases the length of the muscles responsible for the action (called the prime movers or agonists, see Chapter 6) to that at which maximum tension is developed.
- It provides neural facilitation (through the Golgi tendon organs) by contraction of the group of muscles that may slow the movement in the recovery phase (the antagonists).

The important functions of the action phase are:

- Movements are initiated by the large muscle groups and continued by faster, smaller, more distal muscles of the limbs, increasing the speed throughout the movement as the segmental ranges of movement increase (see Appendix 2.1).
- The accuracy of the movement increases through the recruitment of muscles with progressively decreasing innervation ratio (see Appendix 2.1).
- The segmental forces are applied in the direction of movement and movements are initiated with minimum inertia as movement proceeds along the segment chain (see Appendix 2.1).

The recovery phase:

- Involves controlled deceleration of the movement by, usually, eccentric (lengthening) contraction of the appropriate antagonist muscles.
- May achieve a position of temporary balance, as at the end of a golf swing.
- For a learner, may require a conscious effort to overcome neural inhibition from the Golgi tendon organs, which is reinforced by antagonistic muscle spindle activity.

Although phase analysis is very helpful in identifying critical features of a movement and in building a deterministic model, it does have important limitations. The phase boundaries need to be easily observable, but there is some arbitrariness in specifying them, as in the tennis serve example above. The end of the backswing of the racket might conveniently define transition from the preparation to the action phase; however at that instant, the legs and trunk will be in their 'action' phase while the distal joints of the racket arm may still be in their 'preparation' phase. The extension of the three-phase ballistic model to other activities is unhelpful, although common. The biomechanical structure of other movements may require more than three phases, as in the following examples.

Phase analysis of running

In Chapter 1, we divided the running cycle into two rather obvious phases: stance and swing, for which the key events are clearly touchdown and toe-off. The functions of these two phases can then be defined as follows. The stance phase absorbs impact, allows an active landing, maintains forward momentum, and accelerates to toe-off. The swing phase allows recovery of the non-support leg and prepares for foot descent to hit the ground with an active landing. This two-phase schema results in overly complex biomechanics of the phases. An alternative, more attractive schema biomechanically, subdivides the movement into three phases each within the above stance and swing phases, as follows. Although this phase analysis is far more biomechanically sound, it is unlikely that the phase boundaries of the short duration foot-strike phase will be discernible from 50 or 60 Hz digital video recordings, an important practical drawback.

- Foot strike – starts at first foot contact (touchdown). This is a very short phase in which the foot should, ideally, hit the ground moving backwards. The main function of this phase is impact absorption.
- Mid-support – starts after the impact peak as limb length starts to shorten. This phase is characterised by relative shortening of the overall length of the support limb towards the centre of mass. The hips and knees flex (see also Figure 3.13(a), region a–b), the ankle dorsiflexes and the foot pronates. Its functions are maintenance of forward momentum, support of body weight and storage of elastic energy.
- Take-off – starts as limb length starts to increase, ends at toe-off. This phase involves relative lengthening of the overall limb length. The hips and knees extend (see also Figure 3.13(a), region c–d), the ankle plantar flexes and the foot supinates. Its functions are to accelerate the body forwards and upwards and to transfer energy from the powerful thigh muscles to the faster muscles of the calf.
- Follow-through – starts at toe-off. This phase is characterised by slowing of hip extension then start of hip flexion, both accompanied by, and the latter assisting, knee flexion (see also Figure 3.13(a), region d–e). It functions as a decelerating phase for the recovery leg.
- Forward swing – starts as the foot begins to move forward relative to the body. The hip flexes with the knee flexing or flexed (see also Figure 3.13(a), at f). Its main function is to coordinate the recovery leg movement to enhance the forwards–upwards reaction force from the ground, as this phase coincides with the take-off phase of the other leg (similar to the coordination of arm and leg movements in the vertical jump examples in Chapter 1). The phase also prepares for the next, foot descent, phase.

- Foot descent – starts as the forward movement of the leg and foot stops, ends at first foot contact. The hip and knee extend (see also Figure 3.13(a), region g–a) after the arresting of the forward leg swing by the hamstring muscles; the gastrocnemius muscle at the back of the calf is pre-activated for landing. This phase's main function is to have the foot strike the ground in an active landing, to prevent braking of forward movement of the centre of mass.

Phase analysis of more complex movements

Three examples are given below, focusing on the name and main functions of each phase only.

Phase analysis of volleyball spike

This is probably best analysed in seven phases:

1 Run-in: generating controllable speed.
2 Landing: impact absorption.
3 Impulse drive: horizontal to vertical momentum transfer.
4 Airborne phase of preparation: as ballistic.
5 Hitting phase: as ballistic.
6 Airborne phase to landing – airborne recovery: as ballistic.
7 Landing: to absorb impact, control deceleration and prepare for next move.

Phases 4 to 6 are seen to follow the ballistic sequence within a more complex movement.

Phase analysis of javelin throw

1 Run-up: generating controllable speed.
2 Crossover steps: 'withdrawal' of javelin to extend acceleration path; transfer from forwards to sideways action.
3 Delivery stride – the action phase; similar, in many ways, to that for ballistic movements.
4 Recovery – to avoid crossing foul line.

Historically, phase 2 would have been split into the withdrawal and one crossover step. Phases 2 to 4 follow the ballistic sequence within a more complex movement structure.

Phase analysis of front crawl swimming

Swimming can, conveniently, be broken down into the time spent starting, stroking and turning. Focusing only on the stroking part of this subdivision for the front crawl, we can identify the following phases for each stroke:

1   Initial 'press' – smooth entry of the hand into the water.
2   Outward scull – propulsive force generation by the arm moving outwards and backwards through the water; this extends the acceleration path; the hand acts as a hydrofoil.
3   Inward scull – propulsive force generation by the arm moving inwards and backwards through the water; this extends the acceleration path; the hand acts as a hydrofoil.
4   Recovery – moving the arm through the air for the next stroke.

3 More on movement patterns – the geometry of motion

Knowledge assumed
Ability to undertake simple
analysis of videos of sports
movements (Chapter 1)
Familiarity with graphs of one
variable plotted as a function
of a second variable, often
time

INTRODUCTION

In Chapter 1, we considered human motion in sport and exercise as 'patterns of movement'. This chapter is designed to extend your understanding of the 'geometry' of movement in sport and exercise. We will outline some questions that determine which type of movement pattern may be of most use to a movement analyst. The branch of biomechanics that deals with the geometry of movement without reference to the forces causing the movement is often called 'kinematics', although we will generally avoid this technical term.

Much of our work as sports biomechanists involves the study and evaluation of how sports skills are performed. To analyse the observed movement 'technique', we need to identify important features of the technique. For a qualitative biomechanical analyst, this means being able to observe those features of the movement; for the quantitative analyst, this requires measuring those features and often, further mathematical analysis. As I explained in Chapter 1, the increasing demand from the real world of sport and exercise – coaches, athletes and other practitioners – outside of academia has generated an increasing demand for good qualitative movement analysts; this is again our main focus in this chapter.

BOX 3.1 LEARNING OUTCOMES

After reading this chapter you should be able to:

- explain the various forms of movement pattern that are important for any movement analyst
- be able to interpret graphical patterns of one movement variable (position, velocity or acceleration) in terms of the other two from the shapes of the patterns
- define linear, rotational and general motion
- understand the physical model appropriate to each type of motion and appreciate the uses and limitations of each model
- appreciate the importance of linear and angular motion in sport and exercise
- understand the basic movement patterns used to study joint coordination.

MOVEMENT PATTERNS REVISITED

In Chapter 1, we focused on video sequences as the basic representation of movement patterns in sport. We noted, however, that these sequences are complex, because they contain so much information. It is often beneficial for the qualitative or quantitative movement analyst to look at simpler, if less familiar, representations of movement patterns. In this chapter, we will explore the various patterns of movement that a movement analyst can use, at his or her discretion, to supplement (although not to replace), qualitative video analysis. They are all, even if this is not immediately obvious to you at this stage because of their unfamiliarity, far simpler than video recordings because they focus on aspects of a movement pattern that a movement analyst needs to observe and analyse; this simplicity underpins their usefulness to movement analysts.

Figure 3.1 Stick figure sequences of skier: (a) front view; (b) side view; (c) top view.

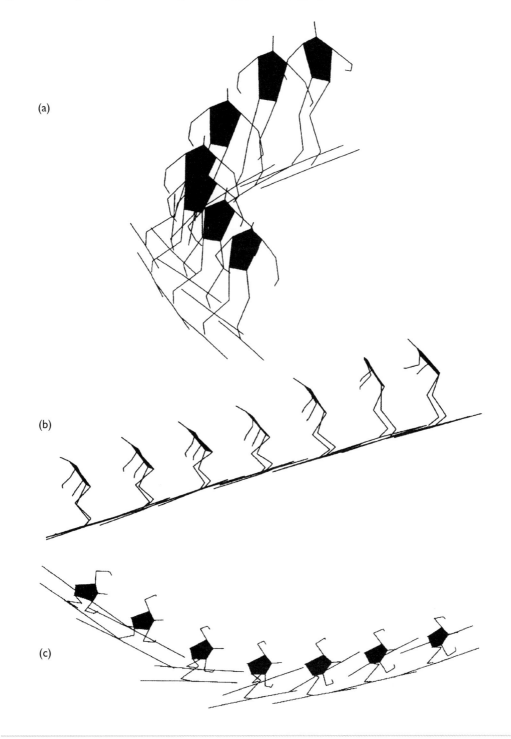

These 'movement analysis supplements' are particularly relevant when we ask the question that kept arising in the section on 'Fundamental movements' in Chapter 1, 'How are these movements coordinated to produce the desired outcome?' They may be described, for now, as computer-generated animation sequences and graphs.

Computer-generated animation sequences are simplified representations of video movement sequences and are obtained from them. The degree of 'abstraction' can run from 'skeletal' stick figures (Figure 3.1) to 'solid body' representations (such as that of Figure 3.2) to ones that can have almost as much detail as the video sequence itself.

Graphs can be of several types, of which the most useful for the movement analyst are time series, angle–angle diagrams and phase planes. Time series are simply graphs (or plots), of one movement variable, such as joint angle, as it changes over time during the course of the movement. Several time series can be plotted on the same graph. Angle–angle diagrams are graphs of one joint angle as a function of another. The focus is how one angle changes with changes in a second angle; in other words we focus on how the two angles 'co-vary' rather than how they each evolve with time. Angle–angle diagrams are used extensively in the study of movement coordination. Phase planes are normally graphs of the angular velocity of one joint as a function of the angle of that same joint. The focus is on the so-called 'coordination dynamics' of that joint. Phase planes are also used extensively in the study of movement coordination. All three of these graphs will be discussed in detail in the following sections.

Figure 3.2 Solid body model of cricket fast bowler.

FUNDAMENTALS OF MOVEMENT

Motion in sport can be linear, rotational or, more generally, a combination of both. Each type of motion is associated with a physical model of the sports performer or sports object being studied. The behaviour of this physical model can be represented by mathematical equations, which constitute a 'mathematical model' of the object or performer. These mathematical models can then be used to investigate how the object or performer moves.

Linear motion and the centre of mass – the 'point' model

Linear motion is movement in which all parts of the body travel the same distance in the same time in the same direction. If two points on the body are joined by a straight line then, in successive positions, this straight line will remain parallel to its initial orientation. Linear motion is often subdivided into rectilinear motion, which is one-dimensional (as for a puck travelling across an ice rink), or curvilinear, which can be two- or three-dimensional (as for the centre of mass of a shot when in the air or the centre of mass of a ski jumper). It is important to note that, although in curvilinear motion the centre of mass – like the skis in Figure 3.3 – follows a curved path, no rotation about the centre of mass is involved.

The overall linear motion of a sports performer or object, such as a shot, can be specified by the motion of a single point. This imaginary point, which has the same mass as the performer or object, is known as the centre of mass. The mass of the

Figure 3.3 Curvilinear motion.

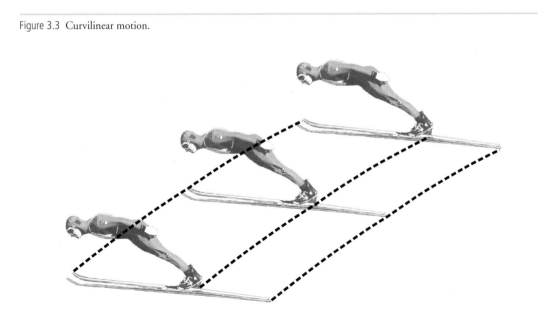

performer is evenly distributed about this point. The motion of the centre of mass of an object, including the sports performer, totally specifies the linear motion of the object as a whole, even if the object's overall motion is more general. Linear motion ignores any rotation about the centre of mass, which means that it only tells us part, often only a very small part, of the movement pattern we are studying. Linear motion is sometimes known as 'translation', but we will use 'linear motion' throughout this book.

Rotation and rigid bodies – the 'rigid body' model

Rotation, or angular motion, is motion in which all parts of an object travel through the same angle in the same time in the same direction about the axis of rotation. The movement of a body segment about a joint, as in Figure 3.4, is motion of this type. An object that retains its geometrical shape, for example a cricket bat, can be studied in this

Figure 3.4 Angular motion.

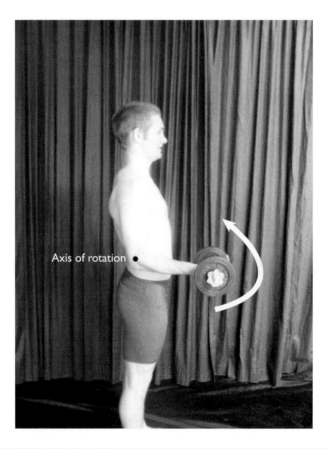

way and is known as a rigid body. Human body segments, such as the thigh and forearm, are often considered to be close enough approximations to rigid bodies to be treated as such. The complete human body is not a rigid body, but there are sports techniques, as in diving and gymnastics, in which a body position is held temporarily, as in a tuck or pike. During that period, the performer will behave as if he or she were a rigid body and can be described as a 'quasi-rigid' body. In most sport activities, the human performer does not behave as one rigid body. We then need to use the following model.

General motion and the human performer – the 'multi-segmental' model

Most human movements involve combinations of rotation and linear motion, for example the cross-country skier shown in Figure 3.5.

Such complex motions can be represented by a linked, multi-segment model of the performer, where each of the body segments is treated as a rigid body. The rigid bodies are connected at the joints between body segments. Such movement patterns can be very difficult to analyse, but simpler representations of the structure of the movement can be very helpful, as we shall see below.

Figure 3.5 General motion.

LINEAR MOTION AND THE CENTRE OF MASS

A sprint coach might want to know how long it takes one of his or her novice athletes to reach maximum horizontal velocity, often referred to as running speed, and what happens thereafter, when running a 100-m race. To do this, we would need to record the position of the athlete's centre of mass (we will return to how we can do this in Chapter 5) at equal intervals of time – we can do the latter using a video camera. Let us assume for now that the movements of a particular point that we mark on the pelvis closely approximate those of the centre of mass; we could obtain the positions of this point from biomechanical software packages for qualitative video analysis, such as siliconCOACH™ (siliconCOACH Ltd, Dunedin, New Zealand; http://www.siliconcoach.com) and Dartfish™ (Dartfish, Fribourg, Switzerland; http://www.dartfish.com). This example is a curvilinear movement, but we make it rectilinear by ignoring the vertical movements of the sprinter and focusing only on the horizontal ones. Our geometrical pattern in this case represents the horizontal path taken by the centre of mass during the time of the race and might look something like Figure 3.6, a pattern of the centre of mass movement – its horizontal displacement – over time.

It is often said that a picture is worth a thousand words, so let us see what we can deduce from the movement pattern of Figure 3.6, which, as we saw above, is known as a 'time series', because we are looking at the pattern over time. But first, a few important terms.

The time-series graph of Figure 3.6, which is a special kind of pattern, is called a displacement–time graph – this, for all important purposes, is the same as a position–time graph. The rate of change of displacement or position with time is known as velocity; the rate of change of velocity with time is called acceleration. We can find important information about velocities and accelerations from a displacement–time graph qualitatively, without recourse to any mathematics, if we accept two extremely important relationships:

Figure 3.6 Hypothetical horizontal displacement of the centre of mass with time for a novice sprinter.

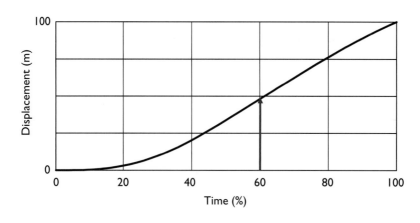

- The velocity can be obtained from the gradient – or slope – of a displacement–time graph. Likewise acceleration can be obtained from the slope of a velocity–time graph if we have one.
- The acceleration can be obtained from the curvature of a displacement–time graph.

So, from our displacement–time graph, we can qualitatively describe two further movement patterns; we can specify important features of the velocity–time pattern and the acceleration–time pattern. As we will now see, this qualitative information will answer fully the questions that our hypothetical sprint coach posed above.

The gradient – or slope – of a line should be intuitive. All you need to do is to consider 'walking' along the graph from left to right:

- If you are going uphill, the gradient, that is the velocity, is positive. This happens throughout Figure 3.6.
- If you are going downhill, the gradient and velocity are negative.
- Where the direction changes from uphill to downhill, the gradient and velocity are zero, and the gradient and velocity change from positive to negative.
- Likewise, where the direction changes from downhill to uphill, the gradient and velocity are zero and the gradient and velocity change from negative to positive.

This doesn't, perhaps, seem to help much with our coach's questions.

Curvature is probably less intuitive, so let us define two types of curvature as in Figure 3.7, which we can call 'valley-type' and 'hill-type' curvature; we will also use the terms 'positive' and 'negative' curvature. Positive curvature is what you would see looking at a simple curved valley from the side and negative curvature corresponds to the side view of a simple curved hill (Figure 3.7). Let us walk along the graph in Figure 3.6 again, seeking this time to indicate curvature not gradient. At the start of the race, the curvature is positive (valley-type); somewhere it changes at about 60% of the race time to negative. And what is the gradient doing there? It changes from becoming steeper to becoming less steep; in other words, the velocity stops increasing and starts decreasing (Figure 3.8). Our coach's novice performer takes 60% of the race to reach maximum speed and then starts to slow down; we have answered our coach's question. This pattern is not, I must admit, typical for a good sprinter, who would probably maintain a constant speed from about 30 m to 90 m, before slowing slightly,

Figure 3.7 Top: positive (valley-type) curvature; bottom: negative (hill-type) curvature.

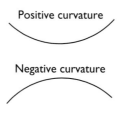

Positive curvature

Negative curvature

but it does provide a simple example of qualitatively interpreting curvature. To summarise (see also Figure 3.7):

- When the curvature is positive, the acceleration is positive, and the gradient and velocity are increasing; this is why we called valley-type curvature positive.
- When the curvature is negative, so is the acceleration, and the gradient and velocity are decreasing; this is why we called hill-type curvature negative.
- Where the curvature and acceleration change from positive to negative, the acceleration is instantaneously zero; the velocity stops increasing and starts decreasing.
- Where the curvature and acceleration change from negative to positive, the acceleration is instantaneously zero; the velocity stops decreasing and starts increasing.

For an elite sprinter, we might find a straight horizontal portion on the displacement–time graph where the curvature was zero. The curvature would change from positive to zero at the start of that portion of the graph and from zero to negative at its end. The acceleration during that portion of the graph would be zero and the gradient, the velocity, would be constant.

When you have the above information about what the changes in gradient and curvature on a displacement–time graph represent for velocity and acceleration then, with experience, you will find that you can roughly – and non-numerically – sketch the velocity and acceleration patterns to obtain further movement patterns for qualitative analysis. All quantitative analysis packages will do this too, and provide more accurate velocity and acceleration patterns; these can also be analysed qualitatively through their shape as well as quantitatively.

Figure 3.8 Hypothetical centre of mass displacement (continuous black curve), velocity (dashed black curve) and acceleration (blue curve) variation with % race time for a novice sprinter.

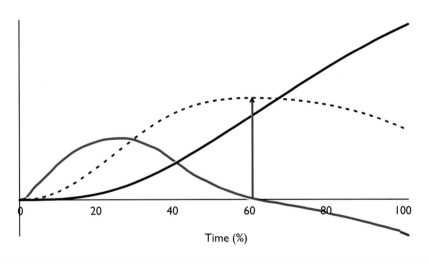

THE GEOMETRY OF ANGULAR MOTION

In the previous section we considered linear motion – the movement patterns of a point. The sports performer or sports object can be represented as a point situated at the centre of mass. The linear motion of this point is defined independently of any rotation taking place around it. The centre of mass generally serves as the best, and sometimes the only, point about which rotations should be considered to occur. All human motion however involves rotation, or angular motion, for example the movement of a body segment about its proximal joint.

Angular motion is far more complex than linear motion. The rules that apply to the rotation of a rigid body can be directly applied to an object such as a cricket bat or, as an approximation, body segments. They can also be applied to a non-rigid body that, instantaneously, is behaving as though it was rigid, such as a diver holding a fully extended body position or a gymnast holding a tuck. Such systems are classified as quasi-rigid bodies. Applications of the laws of angular motion to non-rigid bodies, such as the complicated kinematic chains of segments that are the reality in most human movements, have to be made with considerable care. The theory of the rotation of even rigid bodies in the general case is complicated and many problems in this category have not yet been solved.

The movement variables in angular motion are defined similarly to those for linear motion. Angular displacement is the change in the orientation of a line segment. In two-dimensional motion, also known as planar motion because it takes place in a two-dimensional plane, this will be the angle between the initial and final orientations regardless of the path taken. Angular velocity and acceleration are, respectively, the rates of change with time of angular displacement and angular velocity.

Joint angles are the most important examples of angular motion – first, as here, when we look at the change of the angle over time and then, in this next section, for other combinations of variables. Joint angle patterns are, in general, far more important than linear motion patterns because they open the way to so many fascinating representations of human movement patterns.

The angle–time pattern of Figure 3.9, a time-series pattern, shows how the knee angle changes with time over one stride of treadmill running. As with the linear motion example of Figure 3.6, we can learn a lot about this pattern by studying its geometry – the gradients and curvatures of the graph – as we move through time from left to right. As this analysis focuses on the qualitative aspects rather than the numbers, you need to remember the angle convention that we adopted in Chapter 1: here, a fully extended knee would be a larger angle than with the knee flexed – the angle increases upwards in the figure.

Before you read on, satisfy yourself that you can:

- Identify where the knee is flexing and where it is extending in Figure 3.9(a).
- Determine the gradient of the graph in each region of the overall pattern (Clue – would you be going uphill or downhill walking along the pattern?)

Figure 3.9 Variation of knee angle with time in treadmill running: (a) angle–time series only; (b) regions of flexion and extension (negative and positive gradient); (c) regions of positive and negative curvature.

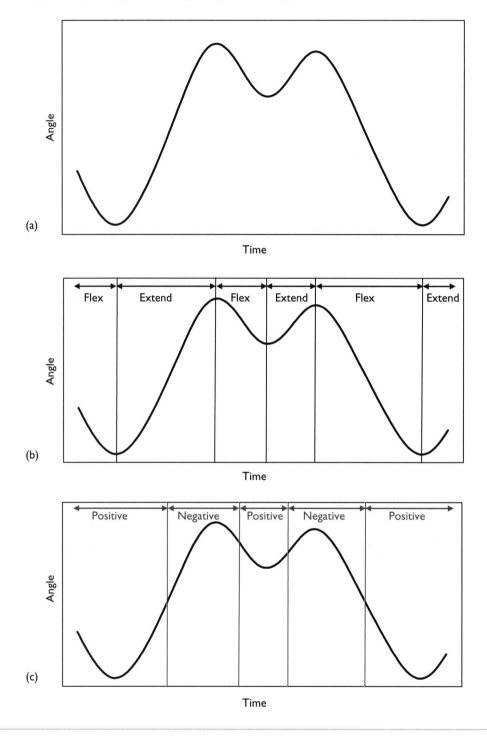

- Identify the curvature of the various regions of the pattern – are they positive (valley-type) or negative (hill-type); refer to Figure 3.7 if you have forgotten.

Check your answers with Figures 3.9(b) and (c). The gradient changes at the vertical black lines in Figure 3.9(b) from positive to negative or from negative to positive; the gradient is instantaneously zero at the vertical lines. The curvature changes, again from positive to negative or vice versa, at the vertical blue lines in Figure 3.9(c); the curvature is instantaneously zero at these vertical lines. The regions of flexion and extension (Figure 3.9(b)), and those of positive and negative curvature (Figure 3.9(c)), are also shown. Study these patterns very carefully and ensure that you understand them fully before carrying on. If you were wrong, go back to Figure 3.9(a) and try to ascertain where and why you went wrong.

Figure 3.10 is a combination of Figures 3.9(b) and (c), to which have been added the angular velocity (continuous blue curve) and acceleration (dashed blue curve) patterns. Angular velocity – the chain-dotted pattern – is positive when the knee is extending and negative when it is flexing. The horizontal line marked 0 shows where the angular velocity or acceleration is zero; values below this line are negative, those above are positive. The angular acceleration – the continuous pattern – is positive, corresponding to positive (valley-type) curvature of the knee angle curve, when it is driving the knee from a flexed to an extended position; we can call this an extending acceleration. The angular acceleration is negative, corresponding to negative (hill-type) curvature of the

Figure 3.10 Variation of knee angle (black curve), angular velocity (continuous blue curve) and angular acceleration (dashed blue curve) with time in treadmill running. Vertical black lines separate positive (extension) and negative (flexion) slope (velocity) and vertical blue lines separate positive and negative curvature (acceleration). The angle, angular velocity and angular acceleration data have been 'normalised' to fit within the range −1 to +1.

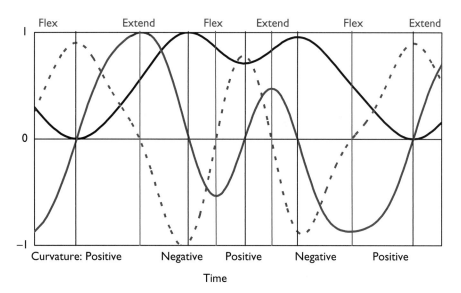

knee angle curve, when it is driving the knee from an extended to a flexed position; we can call this a flexing acceleration. Trace through the entire patterns of Figure 3.10 until you thoroughly understand them and can explain them to your fellow students; this is a very important step towards becoming a good movement analyst.

Another noteworthy geometric feature of Figure 3.10 is the sequence of the three movement patterns: an extending (positive) angular acceleration, caused by muscle tension, occurs before the angular velocity changes from flexing (negative) to extending (positive). The resulting sequence of peaks (or troughs) is: acceleration, velocity, angle. Also notable is the inverse phase relationship between the angle and angular acceleration patterns; one is increasing while the other is decreasing; this is typical of cyclic joint movements, but not always so apparent in movement patterns in discrete sports skills, such as jumping and throwing.

THE COORDINATION OF JOINT ROTATIONS

Before looking at how we interpret graphical representations of coordination, let us begin by considering what we mean by this important term. In Chapter 1, we saw how well-coordinated arm movements can improve the height achieved in a standing vertical jump. So, we could study coordination of arm movements with vertical forces for vertical jumps; this would be an ambitious starting point, however. In Chapter 2, when considering movement principles, one of the universal principles we noted was 'Mastering the many degrees of freedom involved in a movement'. This is one explanation of what coordination involves. A rather longer definition, which elaborates on the one in the previous sentence, introduces the idea of 'coordinative structures'. This viewpoint sees the acquisition of coordination as constraining the degrees of freedom into coordinative structures, which are functional relationships between important anatomical parts of a performer's body, to perform a specific activity. An example would be groups of muscles or joints temporarily functioning as coherent units to achieve a specific goal, such as hitting a ball. As muscles act around joints, this explanation leads us to look at joints and their inter-relationships to gain an initial insight into how sports movements are coordinated. We look at two common ways of doing this in the following subsections.

Angle–angle diagrams

In the previous section, we looked at joint angles as a function of time – a 'time series'. However, times series involving several angles, as in Figure 3.11, can be difficult to interpret for coordination. An alternative is to plot angles against each other – these are called angle–angle diagrams. We could plot three angles in this way to form a three-dimensional plot, but this is rarely done.

Several forms of coordination can be brought to light through angle–angle diagrams.

Figure 3.11 Hip (blue continuous curve), knee (black curve) and ankle (blue dashed curve) angle–time series for three strides of treadmill locomotion: (a) walking; (b) running. Black vertical line indicates touchdown and blue vertical line toe-off.

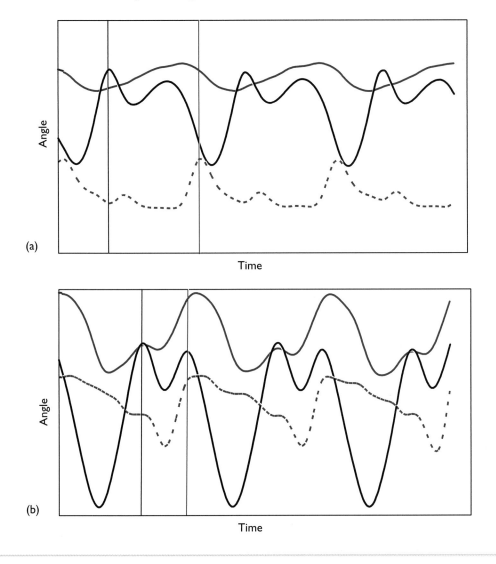

(a)

Time

(b)

Time

The first is 'in-phase' coordination, as when the hips and knees extend during the upward phase of a standing vertical jump. If the two angles change at the same rate, the result is a linear relationship, such as that of Figure 3.12(a). More often, the joints will show in-phase 'turning point' coordination, as in Figure 3.12(b). The second basic form is called 'anti-phase' or 'out-of-phase' coordination; an example is when one joint flexes while the other extends, as in much of the action phase of a kick, in which the hip flexes while the knee extends. Linear 'anti-phase' coordination is shown in Figure 3.12(c) and anti-phase turning-point coordination in Figure 3.12(d). Figure

Figure 3.12 Basic types of coordination. In-phase: (a) linear; (b) turning-point coordination. Anti-phase: (c) linear; (d) turning-point coordination. (e) Phase offset or decoupled coordination.

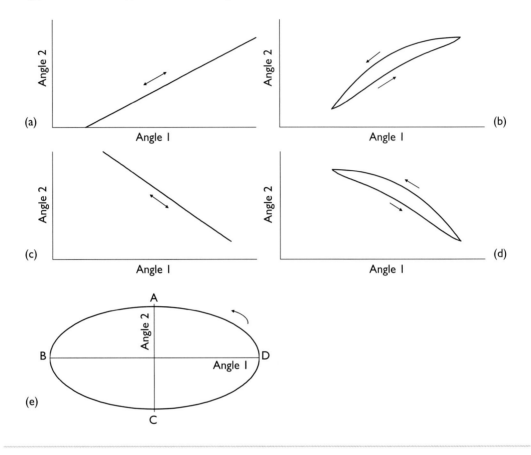

3.12(e) shows 'phase offset' or decoupled coordination. Reading from point 'A' anti-clockwise, both angles flex until 'B', then angle 2 continues to flex while angle 1 extends. From 'C' to 'D', both angles extend and, finally, angle 1 flexes from 'D' to 'A' while angle 2 continues to extend. Coordination in actual human movements is often more complex than these basic patterns, as for the hip–knee angle–angle diagram for one running stride in Figure 3.13. Please note that the diagram has been slightly 'massaged' so that it forms a continuous loop, which is rarely the case owing to movement variability and measurement errors, as shown in Figure 3.14.

Reading around Figure 3.13, starting at the lower right-hand spike at 'a' (which corresponds roughly to touchdown, or heel strike) and progressing anticlockwise, the pattern is as follows. At the start of the stance phase, the hip and knee both flex until 'b'; then, briefly, the knee continues to flex while the hip extends to 'c'. From 'c' to 'd', the two joints extend in phase. From 'd', which is roughly at toe-off, another brief period until 'e' sees the knee flex while the hip extends at the start of the swing phase. From 'e'

Figure 3.13 Angle–angle diagrams for one 'ideal' running stride: (a) hip–knee coupling; (b) ankle–knee coupling; (c) ankle–hip coupling.

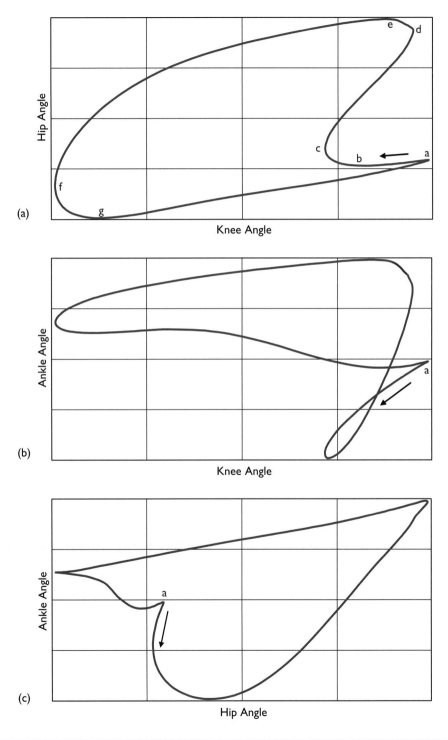

Figure 3.14 Angle–angle diagrams for three strides in treadmill running: (a) hip–knee coupling; (b) ankle–knee coupling.

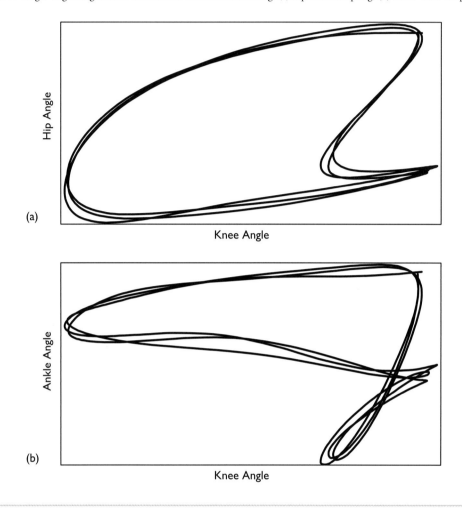

until 'f' both joints flex in-phase during the next part of the swing phase, after which the knee extends while the hip continues to flex until 'g'; both joints then extend in-phase until around touchdown. Note that this pattern involves seven changes in the co-ordination of the two joints. Six of these changes are from in-phase to anti-phase or vice versa, similar to the changes in the simplified pattern known as decoupled coordination (Figure 3.12(e)). Only one change, at point 'a' in Figure 3.13(a) close to touchdown, is from in-phase (extension) to in-phase (flexion). You should now try to repeat this description of joint movements for the same running stride but looking at the ankle–knee and ankle–hip joint couplings, in Figures 3.13(b) and (c) (see Study task 3).

The study of movement coordination is crucial for the movement analyst; one method uses angle–angle diagrams, which have both advantages and disadvantages. Their advantages include that we don't have to flip between angle–time graphs (such

Figure 3.15 Angle–angle diagrams for one walking stride: (a) hip–knee coupling; (b) ankle–knee coupling; (c) ankle–hip coupling.

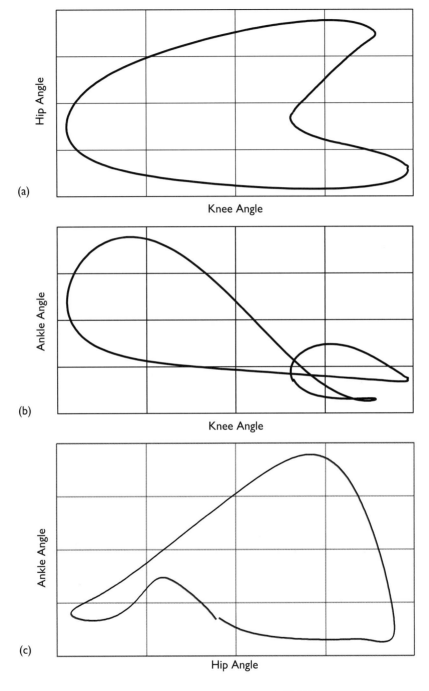

as those of Figure 3.11) and that we can pair joint angles of interest easily to show how they co-vary. These graphs show coordination patterns qualitatively, which can facilitate comparisons, for example between individuals and for one individual during rehabilitation from an injury. We can also compare patterns between, for example, running and walking; most of these comparisons have been based on methods to quantify angle–angle diagrams. Such a reduction of a rich qualitative pattern to a few numbers seems bizarre to me and ignores the saying 'a picture is worth a thousand words'. Few attempts have been made to distinguish patterns qualitatively; one of the very few is known as 'topological equivalence'. Two shapes are topologically equivalent if one can just be stretched – albeit by different amounts in different places – to form the other; two shapes are not topologically equivalent if one has to be 'folded' rather than just stretched to form the other. Simplistically, this means that if the shapes have different numbers of loops, they are not topologically equivalent; they are then qualitatively rather than just quantitatively different, as for the ankle–knee, but not the hip–knee or ankle–hip couplings when comparing the running angle–angle diagrams in Figure 3.13 with those for walking in Figure 3.15. You should also note that the number of changes in the coordination of the two joints during one stride differs between running and walking. For example, in the previous paragraph, we noted seven such changes for the hip–knee coupling in running; Figure 3.15(a) shows only six such changes for walking, all from in-phase to anti-phase or vice versa.

Disadvantages of angle–angle diagrams include their unfamiliarity compared to time series. Also, it is not obvious from the diagram which way round the figure proceeds – clockwise or anticlockwise – or where key events, such as toe-off and touchdown in gait, occur; the latter is also true to some extent for time series. Some criticism can be

Figure 3.16 Angle–angle diagram with time (data) 'points'. The 'distance' on the graph between two successive points shows how much each angle – for the knee the horizontal 'distance', for the hip the vertical 'distance' – changes in one time interval (here 1/50 s).

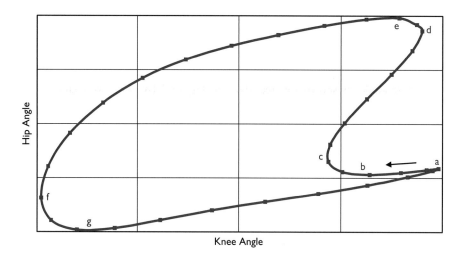

made of the loss of 'time' as a variable, although marking the data, or time, points as in Figure 3.16 adds some indication of how the angles change with time. We do, however, lose access to time-series shape patterns, that is slope = velocity; curvature = acceleration; such relationships do not apply to angle–angle diagrams.

Phase planes

Perhaps the main criticism of angle–angle diagrams is that they do not show coordination changes very clearly without painstaking analysis of the patterns, as for the hip–knee coupling above. Only the ankle–knee coupling of the three joint couplings that we compared between walking and running was qualitatively different topologically between the two modes of gait. Phase planes, a totally different approach, are based on the notion that any system, such as a body segment, can be graphed as diagrams of two variables; for the phase planes used in human movement analysis, these variables are usually joint angle and angular velocity. As it turns out, although the relevance of a phase-plane for a single joint to coordination between joints may seem hard to fathom, phase planes turn out to be pivotal for our understanding of movement coordination, as will be evident later in this section.

Example phase planes, for the hip (Figure 3.17(a)) and the knee (Figure 3.17(b)) joints in a running stride, are shown in Figure 3.17. Its description – although not its analysis – is 'child's play' compared to both angle–time series and angle–angle diagrams. As we define flexion as a decrease in joint angle and extension as an increase, then flexion must be from right to left and extension from left to right in Figure 3.17. Similarly, as we saw in the previous section on time series, a flexion velocity is negative and an extension velocity is positive; so, flexion must be below the horizontal (zero) line in Figure 3.17 and extension above it. And that is about that; the diagrams of Figure 3.17 must progress clockwise. Why? Well let's assume the opposite – they proceed anticlockwise. In this case, in the top half of Figure 3.17(a) the joint would be flexing – from right to left – while the angular velocity was positive, indicating extension. We have proved a contradiction; therefore, our phase planes must progress clockwise with time.

The value of phase planes starts to become evident when we define the so-called 'phase angle' as shown for 'real data' in Figure 3.18. We have changed the graph so that it is 'centred' on its mean value, for reasons that need not concern us in this book. If we now subtract the phase angle – defined anticlockwise from the right horizontal – for one joint from that for a second joint at the same instant, we define a variable known as relative phase. Here, we subtract the knee phase angle from that for the hip, rp = pah – pak. We can do this for every time instant in the cycle to arrive at values of this relative phase as a function of time, which is known as 'continuous relative phase'.

A graph of continuous relative phase as a function of time is shown in Figure 3.19. Continuous relative phase is simply another coordination 'pattern', although more difficult to interpret than angle–angle diagrams. In Figure 3.19, for example, we see that

Figure 3.17 Phase planes for one running stride for: (a) hip joint; (b) knee joint.

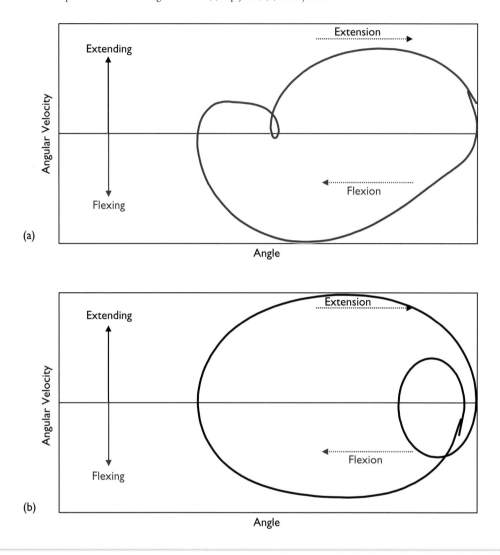

(a)

(b)

the relative phase angle is around 30–60° for much of the swing phase (after toe-off). The value then changes gradually to about 180° at touchdown and fluctuates about that value at the start of the stance phase before gradually returning to around 30–60° before toe-off.

So what, you might be tempted to say! Well, relative phase has been found to be the variable that best expresses coordination changes in a wide range of biological phenomena, including human movement. Examples in human movement include the transitions between walking and running, and bimanual coordination changes. For a further discussion of these and other biological examples of the use of relative phase, see Kelso (1995; Further Reading, page 112).

Figure 3.18 Superimposed phase planes for the hip (blue curve) and knee (black curve) joints in one running stride plus definition of their phase angles (pah and pak) and the relative phase angle (rp).

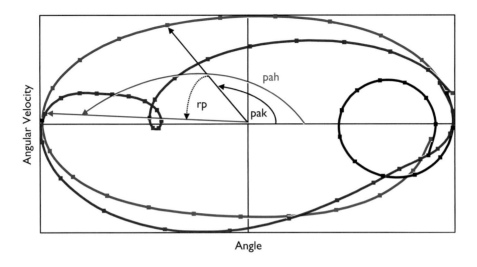

Figure 3.19 Continuous relative phase for hip–knee angle coupling for one running stride, derived from Figure 3.18. The dashed blue line indicates toe-off and the continuous blue line indicates touchdown.

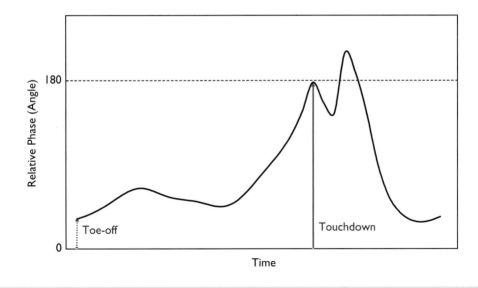

Figure 3.20 Hip and knee phase planes for one stride of walking: (a) hip phase plane; (b) knee phase plane.

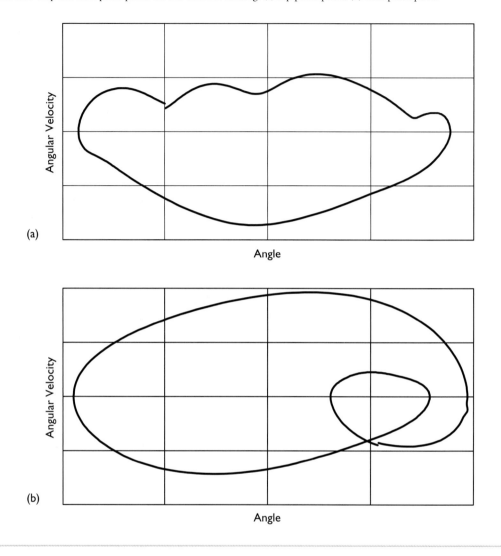

It is also worth noting that phase planes for the hip angle in walking and running are not topologically equivalent (compare Figure 3.20(a) with Figure 3.17(a)); the phase plane for running has two loops while that for walking has only one. However, the knee phase planes for the same running and walking strides are topologically equivalent as they both have two loops, as seen in Figures 3.20(b) and 3.17(b).

BOX 3.2 A CAUTIONARY TALE OF UNRELIABLE DATA

As we have noted many times in this chapter, one of the major tasks for movement analysts is to be able to analyse coordination patterns in sport. Of particular interest to some movement analysts, including me, is the variability that we find in coordination patterns such as angle–angle diagrams, when even a highly skilled performer repeats a movement – this variability is also observed in movement patterns seen as a function of time. Of considerable interest to applied researchers is whether we can accurately and reliably assess such movement variability in competition.

To answer this question, some of my colleagues at the University of Otago and I carried out a study to compare the reliability of estimating movement patterns in laboratory and simulated field conditions. Both conditions were recorded using a digital video camera viewing per-pendicular to the sagittal plane of the runner in our laboratory, with good participant–clothing and clothing–background contrasts. The difference between the two conditions was as follows. In the 'laboratory' condition, the participant had reflective markers attached to his clothing to improve identification of joint axes of rotation; in the 'simulated field' condition, no markers were used because they cannot be in sports competitions. Although this is not the only dif-ference between laboratory and real field conditions, it is usually the most important one by far. The participant ran five trials in each condition at the same speed on a treadmill with equal rest periods between trials. From each trial, we selected three strides, from toe-off to toe-off, to be digitised – this means that a human operator identified each marker on each video frame, effectively manually tracking the markers or estimating the positions of the joints in the no-marker condition. We also tracked the markers automatically using SIMI software (SIMI Reality Motion Systems GmbH, Unterschliessheim, Germany; http://www.simi.com). Five marker trials were auto-tracked using SIMI and manually tracked by four experienced movement analysts on five consecutive days. The four human operators then digitised the five no-marker trials on consecutive days.

Several sources of variability, or variance, are present in this design. The one of interest to movement analysts is movement variability – the variance among trials. The next two sources are due to variability within and among the human operators, respectively known as intra-operator variability and inter-operator variability. The first of these, in our study, was the vari-ance across days and the second that across the four operators. There are other sources of variance from the three two-factor and the one three-factor interactions between our three main factors – trials, days and people. We then used a statistical technique known as analysis of variance (ANOVA) to partition the variances between the three main effects and the several interactions. The results for the marker and the no-marker conditions are summarised in the pie charts of Figures 3.21(a) and (c) for the marker condition and Figures 3.21(b) and (d) for no markers.

Small variances across repeated attempts by the same person (across days in our study) show good intra-operator reliability; small variances across operators (across people) show good inter-operator reliability, sometimes known as objectivity. It should be clear from Figures 3.21(a) and (c) that, when markers are used, movement variability (variance across trials) is by far the

Figure 3.21 Partitioning of variance (as mean squares, MS): (a) individual operator with markers; (b) individual operator without markers; (c) group with markers; (d) group without markers.

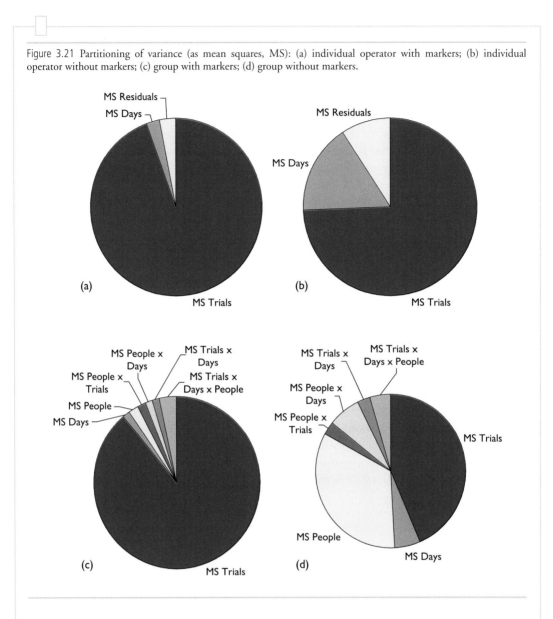

predominant source of variance. Movement variability can, therefore, be assessed both reliably and objectively in these conditions. Indeed, each human operator was not much more inconsistent than auto-tracking, which was 99.99% consistent (and, therefore, reliable). Without markers, however, the picture changed dramatically: true movement variability (across trials) is now obscured by inter-operator (across people, Figure 3.21(d)) and intra-operator (across days, Figures 3.21(b) and (d)) variance. Without markers, therefore, movement variability cannot be assessed reliably or objectively; this is a dramatic finding for applied movement analysts, like me, who have focused much of their research on performance in competition, where markers cannot be attached to the performer.

Although most studies in sports biomechanics have been digitised by only one operator, this practice ignores the importance of assessing objectivity – although, from the results of the study discussed here, this is usually justifiable if markers are used, the same is clearly not the case for no-marker conditions. The results of this study also cast a shadow on previous results from studies in sports biomechanics in which markers have not been used; this applies in particular to those – and there are far too many – in which no attempt has been made to assess reliability or objectivity. Unreliable data is clearly the bane of the quantitative analyst wishing to focus on competition performance; it also presents problems for qualitative analysts whose movement patterns in such conditions will be contaminated by errors.

SUMMARY

This chapter covered the principles of kinematics – geometry of movement – which are important for the study of movement in sport and exercise. Our focus was very strongly on movement patterns and their qualitative interpretation. Several other forms of movement pattern were introduced, explained and explored – including stick figures, time-series graphs, angle–angle diagrams and phase planes. We considered the types of motion and the model appropriate to each. The importance of being able to interpret graphical patterns of linear or angular displacement and to infer from these the geometry of the velocity and acceleration patterns was stressed. We looked at two ways of assessing joint coordination using angle–angle diagrams and, through phase planes, relative phase; we briefly touched on the strengths and weaknesses of these two approaches. Finally, a cautionary tale of unreliable data unfolded as a warning to the analysis of data containing unacceptable measurement errors.

STUDY TASKS

1. Draw sketch diagrams from your own sport and exercise activities to show the three types of motion – linear, angular and general – and the model appropriate to each. List the uses and limitations of each model.

   Hint: You may wish to reread the section on 'Fundamentals of movement' (pages 87–9) before undertaking this task.

2. (a) Download a knee angle–time graph for walking from the book's website. From this, and using the relationships between the gradients and curvatures of the graph, sketch the appropriate angular velocity and acceleration graphs. Remember that you move along the graph from left to right, going uphill and downhill noting the changes in gradient and curvature.

   (b) Repeat the above exercise for the hip angle in walking, which you can download from the book's website.

(c) Keep repeating this for other datasets, until you are able to sketch the angular velocity and acceleration graphs without reference to either the figures in the section on 'The geometry of angular motion' or the answers on the website.

I must stress that analysis of such movement patterns is an essential skill for all movement analysts, whether they approach such a pattern qualitatively or quantitatively. Although these time-series movement patterns are less familiar to you than videos of sports movements, they are far simpler, so persevere with this. Then persevere some more: it will pay great dividends if you become any kind of movement analyst.

Hint: If you are struggling with this task, you may wish to reread the section on 'The geometry of angular motion' (pages 93–6); once you have completed this task, compare your answers with those on the book's website.

3   Describe the coordination sequences of the angle–knee and angle–hip joint couplings in Figures 3.13(b) and (c). Indicate the points on the figures that correspond to coordination changes (similar to points 'a' to 'g' in Figure 3.13(a)). Indicate also whether the joints are in-phase or anti-phase in each region of the diagram (as, for example, for regions 'a' to 'b' and 'g' to 'a' in Figure 3.13(a)) – assume for this purpose that ankle plantar flexion is in-phase with knee and hip extension. Count the number of changes in coordination between the two joints during one running stride; how many of them are from in-phase to anti-phase or vice versa, and how many are from in-phase to in-phase or from anti-phase to anti-phase? Note that the points in Figures 3.13(b) and (c) at which coordination changes will not necessarily be the same as those in Figure 3.13(a); however, point 'a' and an anticlockwise progression are common to all three figures.

Hint: You should study in detail the description of Figure 3.13(a) in the subsection on 'The coordination of joint rotations – angle–angle diagrams' (pages 96–103) before undertaking this task.

For Study tasks 4 to 7, download and save a walking-to-running Excel file from the book's website (for one runner). Successful completion of these four study tasks is absolutely crucial if you want to become a competent movement analyst, so do persevere.

4   By selecting the relevant columns from the Excel file, plot time-series graphs for the hip, knee and ankle angles for both walking and running. Comment on any observable differences between the movement patterns for walking and running.

Hint: You may wish to reread the section on 'The geometry of angular motion' (pages 93–6) and to consult the examples on the book's website before undertaking this task.

5   By selecting the relevant columns, plot angle–angle diagrams for the hip–knee and ankle–knee couplings for each activity. Comment on any observable differences between the coordination patterns for walking and running, such as whether the number of changes in coordination for the same joint coupling during one stride differs between the two forms of locomotion.

Hint: You may wish to reread the subsection on 'The coordination of joint rotations

– angle–angle diagrams' (pages 96–103) and to consult the examples on the book's website before undertaking this task.

6 Again, by selecting the relevant columns, plot phase planes (angular velocity versus angle) for the hip, knee and ankle for each activity. Yet again, comment on any observable differences between the coordination patterns for walking and running. Hint: You may wish to reread the subsection on 'The coordination of joint rotations – phase planes' (pages 103–6) and to consult the examples on the book's website before undertaking this task.

7 Based on the differences you have noted, explain whether each of the three types of movement pattern in Study tasks 4 to 6 help you to identify any qualitative differences between walking and running. Hint: You may wish to reread the sections on 'The geometry of angular motion' (pages 93–6) and 'The coordination of joint rotations' (pages 96–106) before undertaking this task.

8 Have your lecturer arrange, or arrange for yourself, a discussion session to consider the implications of the results of the study summarised in Box 3.2 for qualitative movement analysts. Hint: You should read Box 3.2 several times before undertaking this task.

You should also answer the multiple choice questions for Chapter 3 on the book's website.

GLOSSARY OF IMPORTANT TERMS (compiled by Dr Melanie Bussey)

Angle–angle diagram A graph in which the angle of one joint or body segment is plotted as a function of the angle of another joint or body segment. Conceptually, these can also be three-dimensional – involving three joints.

Geometry Branch of knowledge dealing with spatial relationships.

Gradient The inclination of a line or surface along a given direction.

Kinematics The branch of mechanics that examines the spatial and temporal components of movement without reference to the forces causing the movement.

Movement variability The **variability** that exists within a movement system, which is observable during movement; it is due to non-linear dynamic processes within the movement system.

Phase angle The angle formed between the x-axis of the **phase plane** and the vector of the phase plane trajectory. This angle quantifies where the trajectory is located in the phase plane as time progresses and is used to calculate the relative phase (angle).

Phase plane; phase plot Usually constructed in movement analysis by plotting the angular velocity of a joint or body segment against its angular position. It provides a qualitative picture of the organisation of the system. Conceptually, these can involve any two (or more) properties of a joint or body segment.

Rigid body An idealisation of a body of finite size in which deformation is neglected. In other words, the distance between any two given points of a rigid body remains constant with time regardless of any external forces exerted on it.

Slope The ratio of the 'rise' (change in the *y*-component) to the 'run' (change in the time component) on a variable–time curve. In movement analysis, effectively the same as the **gradient**.

Tangent (line) A line that touches but does not intersect a curved line or surface and that is perpendicular to the radius of curvature of the arc of the curve where the tangent touches the line or surface.

Time series A list of numbers assumed to measure some process sequentially in time.

Variability A measure of statistical dispersion, indicating how the possible values are spread around the expected value. See also **movement variability**.

FURTHER READING

Kelso, J.A.S. (1995) *Dynamic Patterns: The Self-Organization of Brain and Behavior*, Cambridge, MS: MIT Press (Chapter 1: How Nature Handles Complexity). Most of you won't find this plain sailing, but not because of any mathematics – there are no equations in this chapter – but because of the novelty of much of the material. Stick with it; this book has had far more influence than any biomechanics text on the way I now approach the analysis of sports movements. You too might appreciate the genius of Scott Kelso and be inspired by his approach.

APPENDIX 3.1 FURTHER EXPLORATION OF ANGLE–TIME PATTERNS

Here, we find that the points on the angle–time series at which interesting things happened in Figures 3.9 and 3.10 have names. At points A to E in Figure 3.22, the gradient of the tangent to the curve is zero as the tangent is horizontal. The rate of change of angle with respect to time at these points is therefore zero; that is, the angular velocity is zero. The point at which this happens is called a 'stationary point', which may be any one of three types; two of these are evident at A and B in Figure 3.22. These two points are known as 'turning points'. These are stationary points at which the gradient of the angle curve – the angular velocity – changes sign from positive to negative or vice versa. The third point of this type is far more rarely encountered. Points F to I in Figure 3.22 are known as 'points of inflexion', as they are points where the pattern changes its direction of curvature – from positive to negative or vice versa. We return to these later.

Figure 3.22 Variation of knee angle with time in treadmill running; further explanation of angle–time patterns.

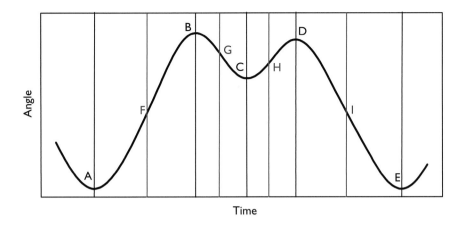

Local minimums

A turning point such as A, C and E in Figure 3.22 is known as a 'local minimum', for which two conditions are necessary:

- The angle at that point must be less than its value at any nearby point; it is not necessarily the overall or 'global minimum' value, hence the use of the term 'local'.
- The gradient of the angle curve, the angular velocity, must change from a negative value (when the angle is decreasing) to a positive value (when the angle is increasing); hence, the angular acceleration is positive.

For an angle–time graph, the above can be summarised as follows: a local minimum angle (A, C or E) corresponds to a condition of zero angular velocity – changing from a negative to a positive value – and a positive acceleration.

Local maximums

A turning point such as B and D in Figure 3.22 is known as a 'local maximum', for which two conditions are necessary:

- The angle at that point must be greater than its value at any nearby point.
- The gradient of the angle curve, the angular velocity, must change from a positive to a negative value; hence the angular acceleration is negative.

For an angle–time graph, a local maximum angle (B or D) again corresponds to a condition of zero angular velocity – changing from positive to negative – but to a negative acceleration.

Points of inflexion

To interpret these, consider the positive and negative curvatures of the angle–time series of Figure 3.22 and the point at which they meet. If the curvature is positive, the gradient of the angle curve and, therefore, the angular velocity, is increasing. The acceleration must, therefore, be positive for this portion of the curve. For an angle–time curve, a positive curvature is a region of positive acceleration, reflecting an increasing angular velocity. If the curvature is negative, so is the acceleration, and the gradient of the angle–time curve – the angular velocity – is decreasing. The points of inflexion, F to I, in Figure 3.22 indicate instances of zero acceleration.

Stationary points of inflexion

A point of inflexion at which the gradient of the tangent happens to be zero fulfils the conditions of a stationary point, which is why it is called a stationary point of inflexion. It does not fulfil the extra condition required for a turning point; that the slope changes sign. These are rarely encountered by human movement analysts.

4 Quantitative analysis of movement

Knowledge assumed
Basic familiarity with use of digital video cameras (Chapter 2)
Position–velocity–acceleration relationships (Chapter 3)
Basic algebra
Qualitative movement analysis (Chapter 2)
Coordination diagrams (Chapter 3)

INTRODUCTION

In this chapter we will mostly explore the use of video analysis – videography – in the study of sports movement patterns, including the equipment and methods used, experimental procedures and data processing. Videography is by far the most likely method of recording movement patterns that an undergraduate student will come across. The increasing computer control of our main data collection equipment in sports biomechanics, along with much more accessible software, has lessened our need for repetitive and tedious calculations, and made mathematical skills less important for many movement analysts. Basic mathematical skills can improve our understanding of sports performance in some cases, two of which are introduced towards the end of this chapter; the first is projectile motion, and the second examines how rotation of a body generates linear velocities and accelerations. Symbolic representations are used in this chapter (some people, mistakenly, call this mathematics), but mathematical derivations are avoided.

BOX 4.1 LEARNING OUTCOMES

After reading this chapter you should be able to:

- understand the importance of videography in the study of sports movements
- undertake a quantitative video analysis of a sports technique of your choice
- understand the important features of video equipment for recording movements in sport
- outline the advantages and limitations of two- and three-dimensional recording of sports movements
- list the possible sources of error in recorded movement data
- describe and implement experimental procedures that would minimise measurement inaccuracy in a study of an essentially two-dimensional movement
- appreciate how these procedures can be extended and modified to record a three-dimensional movement
- understand how 'noise' can be removed from videographic data
- appreciate the need for accurate body segment inertia parameter data, and some ways in which these can be obtained
- explain the differences between vectors and scalars
- calculate the maximum vertical displacement, flight time, range and optimum projection angle of a simple projectile for specified values of the three projection parameters
- understand simple rotation kinematics, including vector multiplication, and the calculation of linear velocities and accelerations caused by rotation.

THE USE OF VIDEOGRAPHY IN RECORDING SPORTS MOVEMENTS

Background

Quantitative biomechanical analysts are mainly interested in improving performance and reducing injury risk. They use a mixture of experimental and theoretical approaches to seek answers to such questions as: What is the best running technique to minimise energy expenditure? How should the sequence of body movements be coordinated in a javelin throw to maximise the distance thrown? Why are lumbar spine injuries so common among fast bowlers in cricket?

As we noted in Chapter 1, we can identify two fundamentally different approaches to experimental movement analysis in sport – qualitative analysis and quantitative analysis; the latter requires detailed measurement and evaluation of the measured data. Earlier chapters in this book had a strong bias towards qualitative analysis whereas this chapter, along with Chapters 5 and 6, will focus mostly on quantitative analysis.

The quantitative experimental approach often takes one of two forms, usually referred to as the cross-sectional and longitudinal approaches. A cross-sectional study, for example, might evaluate a sports movement by comparing the techniques of different sports performers recorded at a particular competition. This can lead to a better overall understanding of the biomechanics of the skill studied and can help diagnose faults in technique. An alternative cross-sectional approach, which is less frequently used, is to compare several trials of the same individual, for example a series of high jumps by one athlete in a competition or in a training session. This is done to identify the performance variables that correlate with success for that athlete. In a longitudinal study, the same person, or group, is analysed over a longer time to improve their performance; this probably involves providing feedback and modifying their movement patterns. Both the cross-sectional and the longitudinal approaches are relevant to the sports biomechanist, although conclusions drawn from a cross-sectional study of several athletes cannot be generalised to a single athlete, or vice versa.

Movement analysts now use single-individual designs far more than in the past, recognising that group designs often obscure differences between individuals in the group and, indeed, the group mean may not apply to any single individual. After all, most athletes are mainly interested in factors that affect their performance or might be an injury risk for them. In a case study, a single person may be analysed on one or just a few occasions; this approach is often used when assessing an injured athlete. A single-individual design usually involves studying that person across time; multiple single-individual designs study individual members of a group of performers across time. This also gives the analyst a chance to use a group design simultaneously with the multiple single-individual study. In such studies, it has been recommended that, for reasonable statistical power, 20 trials per person should be analysed for a group of five performers; for a group of 10 performers, 10 trials each; for a group of 20, five trials each.

To give it a theoretical underpinning, an experimental study should be used in conjunction with a theoretical approach, such as the use of deterministic models or some other way of identifying key factors in the movement (as in Chapter 2), or more advanced movement modelling techniques, such as computer simulation modelling or the use of artificial neural networks, which are beyond the scope of this book.

Videography

The main method currently used for recording and studying sports movements is digital videography. Cinematography – using cine film cameras – is now rarely used, and the same almost applies to the use of 'analog' video cameras; neither will be considered in any detail in this chapter. Motion analysis systems that automatically track skin markers are increasingly used in biomechanics research laboratories; these systems are many times more expensive than video analysis systems, are technically far more complicated, require far more expert operators and currently cannot be used outdoors during daylight hours. For these reasons, which usually mean that students in the earlier years of their study will not encounter such motion analysis systems, they are not dealt with in this book (interested readers should consult Milner (2007); see Further Reading, page 152).

A great strength of videography is that it enables the investigator to record sports movements not only in a controlled laboratory setting, but also in competition. It also minimises any possible interference with the performer. Indeed, performers can be 'videoed' without their knowledge, although this does raise ethical issues that will normally be addressed by your Institutional Research Ethics Committee.

Quantitative analysis will often involve the biomechanist having to digitise a lot of data. This process of 'coordinate digitisation' involves the identification of body landmarks used to aid the estimation of joint axes of rotation. In videography, particularly in three-dimensional studies, this will normally be done by the investigator manually digitising the required points using a computer mouse or similar device. Some video analysis systems can track markers in two dimensions, saving the investigator much time. Automatic marker-tracking systems, as their name implies, track markers automatically, and in three dimensions, although operator intervention may still be needed if too few cameras can see the marker during some part of the movement. Whichever way coordinate digitising is performed, the linear coordinates of each digitised point are recorded and stored in computer memory.

What quantitative analysts measure

After digitising a movement sequence, linear and angular positions and displacements can be calculated and presented as a function of time – a time series (for example, Figure 3.8). Some additional data processing will normally be performed to obtain displacements of the centre of mass of the performer's whole body (Figure 3.5 and Chapter 5). Velocities and accelerations will also probably be obtained from the displacement data (for example, Figures 3.7 and 3.10). As well as these time-series movement patterns, coordination diagrams such as angle–angle diagrams (Figures 3.13 and 3.14) or phase planes (Figure 3.20) can also be obtained; quantitative analysts often want to quantify such graphs, in contrast to the qualitative analyst, whose focus will be on their shape (topography).

Quantitative analysts may also identify values of some variables at important instants in the movement to allow inter- or intra-performer comparisons. These values, often called performance parameters or variables, are usually defined at the key events that separate the phases of sports movements, such as foot strike in running, release of a discus or bar release in gymnastics. They are discrete measures that, although they can be very important for that performer, discard the richness of movement information contained in time-series graphs or coordination diagrams.

Computer visualisation of the movement will also be possible. This can be in the form of stick figure sequences, such as Figures 3.1 and 3.2. These are quick and easy to produce but have ambiguities with respect to whether limbs are in front of or behind the body (Figure 4.1(a)). In three-dimensional analysis, this can be partially overcome by filling in the body and using hidden line removal (Figure 4.1(b)). Full solid-body modelling (Figure 3.2) is even more effective in this respect, but computationally somewhat time-consuming. Solid-body models can also be made more realistic through the use of shading and surface rendering.

Figure 4.1 Computer visualisation: (a) stick figure of hammer thrower; (b) as (a) but with body shading and hidden line removal.

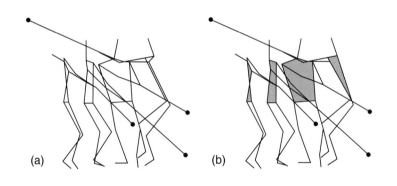

(a)   (b)

Calculating forces and torques (inverse dynamics)

In the previous subsection, we considered 'kinematic' variables, specifically joint angles, angular velocities and angular accelerations. Quantitative analysts also study the 'kinetics' of the movement – forces, torques, and so on. This often involves the calculation of kinetic variables for joints and body segments to try to understand the underlying processes that give rise to the observed movement patterns. The method of 'inverse dynamics' is used to calculate net joint forces and moments from kinematic data, usually in combination with external force measurements from, for example, a force platform (see Chapter 5). The method of inverse dynamics is valuable in sports biomechanics research to provide an insight into the musculoskeletal dynamics that generate the observed characteristics of sports movements; we won't consider this advanced topic further in this book (readers interested in a simple introduction to the topic should see Bartlett, 1999; Further Reading, page 152). Many calculations are needed to determine net joint forces and moments, and an assessment of the measurement and data processing errors involved is important (see Challis, 2007; Further Reading, page 152).

RECORDING THE MOVEMENT

Digital videography

Improvements in video technology in recent years have included the increased availability of electronically-shuttered video cameras. In these cameras, electronic signals are applied to the light sensor to control the time over which the incoming light is detected; this time is usually referred to, somewhat misleadingly, as the 'shutter speed'. It is essential to use electronically-shuttered cameras with a range of shutter speeds to obtain good-quality, unblurred pictures. Although such a range of shutter speeds is fairly normal for a digital video camera (Figure 4.2), the user should ensure that a fast enough shutter speed is used to minimise blurring of the image for the activity being recorded. A setting of 1/1000 s is normally adequate (see also Table 2.1). Be wary of the 'sports' setting on some cheaper digital video cameras, as the shutter speed to which this setting corresponds is often not specified.

Other important developments have included high-quality slow-motion and 'freeze-frame' playback devices that allow the two 'fields' that make up an interlaced video 'frame' to be displayed one after the other. Standard (50 fields or 25 frames per second) video equipment has, therefore, become an attractive alternative to cinematography because of its price, immediacy and accessibility – to such an extent that cinematography has virtually disappeared from sports biomechanics usage. Furthermore, the advent of high-quality digital video cameras has transformed videography. An important reason for this is that digital video can download directly to a computer without the need to record to an intermediate medium, such as a video cassette. Although 'analog' video cameras (which record to VHS – or other format – video cassettes) are still available, they have been largely superseded by digital cameras and will not be considered further, except for comparison with digital video. The development of digital video cameras has meant that a separate video playback system to display images, which was needed with analog video cameras, is no longer necessary as the recording is downloaded onto the analysis computer. The images are stored in the computer and displayed on the computer monitor for digitising.

Figure 4.2 Modern digital video camera.

Video recording is a sampling process; the movement is captured for a short time and then no further changes in the movement are recorded until the next field or frame. The number of such pictures taken per second is called the 'sampling rate' or 'sampling frequency'. For the recording stage of movement analysis, this will correspond to the field rate or frame rate. The overall sampling rate for the analysis may be less if not every field or frame is digitised.

The two main drawbacks of using commonly available digital video cameras for analysing sports movements are the resolution of the image, which restricts digitising accuracy when compared with high-resolution cine digitising tablets, and the sampling rate of 50 fields per second (60 in North America), which makes them unsuitable for the quantitative study of very fast sports movements. For fast movements and for very high-speed events, such as bat–ball impacts, higher frame-rate cameras are needed; such cameras are commercially available but are much more expensive than standard digital video cameras. The 'resolution' of coordinates obtained from digital video images is limited by the number of individual spots of light, or picture elements (pixels), that can be displayed on the computer monitor; 1024 horizontally by 1024 vertically would be a very good specification. This is an important limitation of digital videography, but some video analysis systems use interpolation of the position of the digitising cursor between pixels as a partial solution to this problem.

A further drawback of digital video cameras is that most cannot be 'genlocked' to allow the shutter openings of cameras to be synchronised when using more than one camera to record movement, as in three-dimensional studies. Without a genlock capability, cameras can take pictures up to ½ of a field (or 0.01 s) apart; in quantitative analysis, this discrepancy must be corrected in the analysis software if data accuracy is not to be compromised.

Although digital video cameras are convenient and inexpensive, they often do not allow the interchange of lenses, which can be useful. The quality of lenses on cheaper video cameras may result in image distortion, particularly at wide-angle settings on zoom lenses; this can lead to increased errors in digitised coordinates. Modern, solid-state digital video cameras detect the image using an array of light sensors precisely etched into silicon and are claimed to have zero geometric distortion.

Another important limitation of videography is the vast amount of manual co-ordinate digitisation that is often required for quantitative analysis. This is particularly the case when a three-dimensional analysis is undertaken, as at least two images have to be digitised for each field or frame to be analysed. To overcome this drawback, motion analysis systems such as Vicon™ (Vicon, Oxford, UK; http://www.vicon.com) and EVa RealTime™ (Motion Analysis Corporation, Santa Rosa, USA; http://www.motionanalysis.com), have been developed that automatically track markers attached to the body. Some video software analysis packages, such as SIMI™ (SIMI Reality Motion Systems GmbH, Unterschleissheim, Germany; http://www.simi.com) and APAS™ (Ariel Dynamics, Trabuco Canyon, USA; http://www.arielnet.com) also have an automatic marker-tracking option, but this is currently restricted to two-dimensional movements.

BOX 4.2 TWO-DIMENSIONAL OR THREE-DIMENSIONAL ANALYSIS?

An early decision that must be made in any quantitative video analysis of sports movements is whether a two- or three-dimensional analysis is required. Both have advantages and disadvantages, as summarised by the following.

Two-dimensional recording and analysis:

- Is simpler and cheaper as fewer cameras and other equipment are needed.
- Requires movements to be in a pre-selected movement plane (the plane of motion or plane of performance). It can yield acceptable results for essentially planar movements but it ignores movements out of the chosen plane. This can be important even for an event that might appear essentially two-dimensional, such as the long jump.
- Is conceptually easier to relate to.
- Requires less digitising time and has fewer methodological problems, such as the transformation of coordinates from the video image to the 'real world' movement plane.

Three-dimensional recording and analysis:

- Has more complex experimental procedures.
- Can show the body's true three-dimensional movements.

- Requires less digitising time and has fewer methodological problems, such as the transformation of coordinates from the video image to the 'real world' movement plane.

  Three-dimensional recording and analysis:

- Has more complex experimental procedures.
- Can show the body's true three-dimensional movements.
- Requires more equipment and is, therefore, more expensive. Although it is possible by intelligent placement of mirrors to record several images on one camera, this is rarely practical in sports movements.
- Has increased computational complexity associated with the reconstruction of the three-dimensional movement–space coordinates from the video images, and requires software time synchronisation of the data from cameras that are not physically time-synchronised, as for most digital video cameras.
- Allows angles between body segments to be calculated accurately, without viewing distortions. It also allows the calculation of other angles that cannot, in many cases, be easily obtained from a single camera view. One example is the horizontal plane angle between the line joining the hip joints and the line joining the shoulder joints, which can be visualised from above even if the two cameras were horizontal.
- Raises the problem of which convention to use for segment orientation angles, which two-dimensional analysis sidesteps.
- Enables the reconstruction of simulated views of the performance (e.g. Figures 3.1(a) to (c)) other than those seen by the cameras, an extremely useful aid to movement analysis and evaluation.

Problems and sources of error in motion recording

The recording of human movement in sport can be formally stated as: to obtain a record that will enable the accurate measurement of the position of the centre of rotation of each of the moving body segments and of the time lapse between successive pictures. The following problems and sources of error can be identified in two-dimensional videography of sports movements:

- The three-dimensionality of the position of joint centres of rotation requires the two-dimensional analysis of movements recorded from one camera to be done with care.
- Any non-coincidence of the movement plane (the plane of performance) and the plane perpendicular to the optical axis of the camera (the photographic plane) is a source of error if calibration is performed with a simple scaling object in the plane of motion.
- Perspective and parallax errors need attention. Perspective error is the apparent discrepancy in length between two objects of equal length, such as left and right

limbs, when one of the limbs is closer to the camera than the other. It occurs for movements away from the photographic plane. The term is also sometimes used to refer to the error in recorded length for a limb or body segment that is at an angle to the photographic plane and, therefore, appears to be shorter than it really is. Associated with this error is that caused by viewing away from the optical axis such that, across the plane of motion, the view is not always side-on, as at the positions marked (*) in Figure 4.3. This is sometimes referred to as parallax error. The combined result of these optical errors is that limbs nearer to the camera appear bigger and appear to travel further than those further away. This causes errors in the digitised coordinates.

- Lens distortions may be a source of error, particularly at wide-angle settings on inexpensive zoom lenses.
- Locations of joint axes of rotation are only estimates, based on the positions of superficial skin markers or identification of anatomical landmarks. Use of skin markers can not only help but also hinder the movement analyst, as these markers move with respect to the underlying bone and to one another. The digitising of such markers, or estimating the positions of axes of rotation without their use, is probably the major source of random error (or noise) in recorded joint coordinates. A study that highlighted this point was considered at the end of Chapter 3 (pages 107–9). Locating joint axes of rotation is particularly difficult when the joint is obscured by other body parts or by clothing.
- Other possible sources of error include blurring of the image, camera vibration, digitising errors – related to coordinate resolution and human digitiser errors – and computer round-off errors.

Figure 4.3 Errors from viewing movements away from the photographic plane and optical axis of the camera: (a) side view; (b) as seen from above.

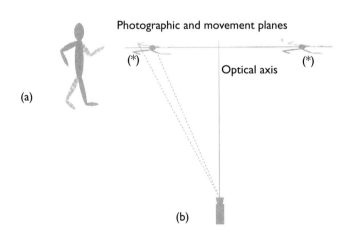

Figure 4.4 A typical calibration object for three-dimensional videography.

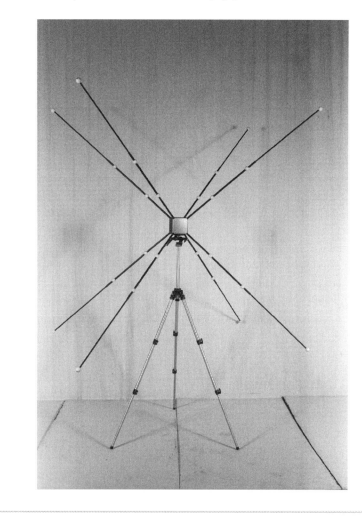

For three-dimensional analysis, there are other potential error sources, although several of those above are partly or wholly overcome.

- Relating the two-dimensional video image coordinates to the three-dimensional movement–space ('real world') coordinates may be a source of error. Several methods of doing this will be considered below but they all present problems. Use of an array of calibration points, such as a calibration object (Figure 4.4) is probably the most common method. Errors within the calibration volume can be accurately assessed, while those outside that volume will be larger and more difficult to assess. Errors will increase with the ratio of the size of the movement space to that of the image.

- All the calibration points must be clearly visible on the images from both cameras; they must also have three-dimensional coordinates that are accurately known.
- Placements of cameras must relate to the algorithm chosen for reconstruction of the movement–space coordinates. Deviations from these requirements will cause errors.

In summary, digitised coordinate data will be contaminated with measurement inaccuracies or errors. These will be random (noise), systematic, or both. All obvious systematic errors, such as those caused by lens distortion and errors in calibration objects, should be identified and removed, for example by calibration or software corrections. Any remaining sources of systematic error will then be very small or of low frequency and will, therefore, have little effect on velocities and accelerations. The remaining random noise in the displacement data, expressed as relative errors, has been estimated as within 1% for a point in the photographic plane for two-dimensional videography and within 2% for a point in the calibration volume for three-dimensional videography. Random errors must be minimised at source by good experimental procedures. Any remaining noise should be removed, as far as possible, from the digitised data before further data processing. These two aspects will be covered in the next two sections. Also, consideration needs to be given to the estimation of errors in digitised coordinate data and their effects on derived values.

EXPERIMENTAL PROCEDURES

Two-dimensional recording procedures

The following steps have often been considered necessary to minimise errors recorded during two-dimensional videography, thereby improving the accuracy of all derived data, and can still often prove useful, particularly for students who are unfamiliar with videography. These procedures also allow a simple linear transformation from image to movement-plane coordinates using simple scale information recorded in the field of view.

- The camera should be mounted on a stationary, rigid tripod pointing towards the centre of the plane of motion and there should be no movement (panning or tilting) of the camera.
- The camera should be sited as far from the action as possible to reduce perspective error. A telephoto zoom lens should be used to bring the performer's image to the required size.
- The focal length of the lens should be carefully adjusted to focus the image. This is best done, for zoom lenses, by zooming in on the performer, focusing and then zooming out to the required field of view. The field of view should be adjusted to coincide with the performance area that is to be recorded. This maximises the size of the performer on the projected image and increases the accuracy of digitising.

- Once the field of view has been set, the focal length of the lens must be kept constant – this means that the auto-focus option available on most digital video cameras must be switched off.
- For particularly long movements, such as the long and triple jumps, consideration should be given to using two or more cameras to cover the filming area (Figure 4.5); with digital video cameras, which do not allow genlocking, another method for event and time synchronisation will then be needed, as described in the next section.
- The movement plane should be perpendicular to the optical axis of the camera. This requires careful attention and can be done in various ways, including the use of laser levelling devices, spirit levels, plumb lines and right-angled (3–4–5) triangles. If the movement takes place in a vertical movement plane on a level horizontal surface, this requirement means that the optical axis of the camera must be both horizontal and at 90° to the vertical movement plane.
- Horizontal and vertical length scales – two 1-m rules are often adequate – and a vertical reference, for example a plumb line, must be included in the field of view. The length scales must be positioned in the plane of motion. Their lengths should be at least that required to give a scaling error, when digitised, of no more than 0.5% of the field of view. Other means of performing scaling include the use of objects resembling chequerboards – if large enough, these would allow calibration and accuracy checks across the whole field of view. This is rarely practical if filming in sports competitions, and is unnecessary providing strict experimental procedures are adopted throughout.
- The background should be as uncluttered as possible, plain and non-reflective. In a laboratory study, contrasting performer and background colours can be helpful.
- The use of colour contrast markers on the appropriate body landmarks is sometimes recommended (see also Box 6.2). These correspond to an axis through the appropriate joint centre, as seen from the camera, when the performer is standing in a specific position and posture: these markers are sometimes referred to as 'joint centre' markers. The use of such markers is not normally possible in competition, and has some potential disadvantages when digitising (see below). If using skin markers, then marking the skin directly is often less problematical than using adhesive markers, which can fall off. Very small markers consisting of a miniature light and battery usually show up very well on video, but like other adhesive markers can fall off.
- To reduce the risk of marker detachment, and to help cope with segmental movement out of the photographic plane, bands can be taped around a segment at the landmarks used to identify the 'joint centre'. This works well for the wrist, elbow, knee and ankle joints but is not possible for some other joints, such as the shoulder and hip.
- If axes of rotation are being estimated from anatomical landmarks, skin markers are not essential, although their use probably saves time when digitising. For standard placement of markers for a two-dimensional study, see Table 5.1 and Box 6.2.
- A sufficiently high sampling rate should be used; 25 Hz is usually adequate for swimming; 50 Hz is often adequate for activities such as the tennis serve, if the

impact of the ball and racket is not the focus of the study; 100 Hz is often needed for quantitative analysis of activities as fast as a golf swing – this is double the maximum sampling rate of most standard digital video cameras. The Nyquist sampling theorem requires that the sampling frequency is at least twice the maximum frequency in the signal (not twice the maximum frequency of interest) to avoid aliasing (Figure 4.6). Aliasing is a phenomenon seen in films when wheels on cars and stagecoaches, for example, appear to revolve backwards. Furthermore, the temporal resolution – the inverse of the sampling rate – improves the precision of both the displacement data and their time derivatives. For accurate time measurements, or to reduce errors in velocities and accelerations, higher frame rates may be needed.

- If several lighting conditions are available, then natural daylight is usually preferable. If artificial lighting is used, floodlights mounted with one near the optical axis of the camera and one to each side at 30° to the plane of motion give good illumination. Careful attention must also be given to lighting when choosing the camera shutter speed.

- Whenever possible, information should be incorporated within the camera's field of view, identifying important features such as the name of the performer and date. The 'take number' is especially important when videography is used in conjunction with other data acquisition methods, such as force plates (Chapter 5) or electromyography (Chapter 6).

- The recording of the movement should be as unobtrusive as possible. The performer may need to become accustomed to performing in front of a camera in an experimental context. The number of experimenters should be kept to the bare minimum in such studies.

- In controlled studies, away from competition, as little clothing as possible should be worn by the performers to minimise errors in locating body landmarks, providing that this does not affect their performance.

- In any videography study, written informed consent should be obtained from all participants; in sport, the coach may be able to provide consent for his or her athletes, but this varies from country to country and should always be checked. Furthermore, approval from your Institutional Research Ethics Committee may be required.

The above procedures impose some unnecessary restrictions on two-dimensional videography, leading to severe practical limitations on camera placements, particularly in sports competitions, because of the requirement that the optical axis of the camera is perpendicular to the movement plane. A more flexible camera placement can be obtained, for example by the use of a two-dimensional version of the direct linear transformation (see below). This overcomes some of the camera location problems that arise in competition because of spectators, officials and advertising hoardings. It requires the use of a more complex transformation from image to movement plane coordinates. Many of the above procedural steps then become redundant, but others, similar to those used in three-dimensional analysis, are introduced. This more flexible

Figure 4.5 Possible camera placements for movement such as long jump.

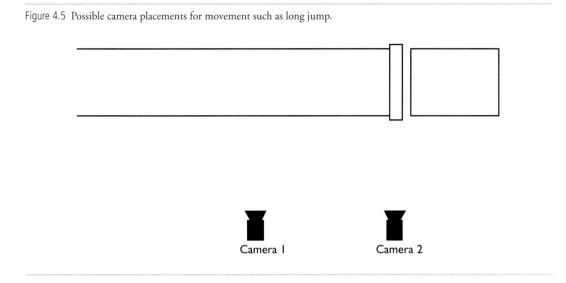

Camera 1   Camera 2

Figure 4.6 Aliasing: (a) the blue signal sampled at the dashed vertical lines is aliased to the black signal, as the sampling frequency is too low; (b) the sampling frequency is sufficiently high to reproduce the correct signal, shown in black.

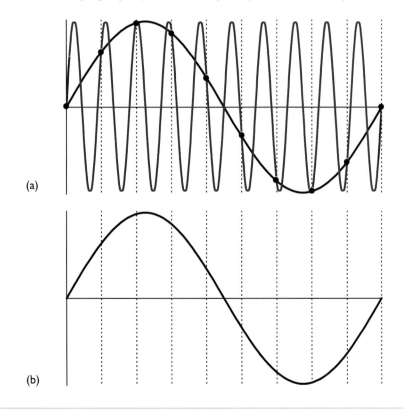

(a)

(b)

approach can also be extended to allow for camera panning. You should assess (see Study task 4) which of the steps in the above procedures could not be ensured if filming at an elite-standard sports competition.

When digitising, attention should be paid to the following points:

- If joint centre markers are used, careful attention must be paid to their movement relative to underlying bones. If segments move in, or are seen in, a plane other than that for which the markers were placed, the marker will no longer lie along the axis through the joint as seen by the camera. To minimise errors, you need a thorough anatomical knowledge of the joints and the location of their axes of rotation with respect to superficial landmarks throughout the range of segmental orientations. It should also be noted that many joint axes of flexion–extension are not exactly perpendicular to the sagittal plane often filmed for two-dimensional analysis.
- The alignment and scaling of the projected image must be checked; independent horizontal and vertical scalings must be performed.
- At least one recorded sequence should be digitised several times to check on operator reliability (consistency). It is also recommended that more than one person digitise a complete sequence to check on operator objectivity.
- The validity of the digitising can only be established by frequent checks on digitiser accuracy, by careful adherence to good experimental protocols, by analysis of a relevant and standard criterion sequence and by carrying out checks on all calculations. Video recording of a falling object is sometimes used as a criterion sequence because the object's acceleration is known.
- Projected image movement may occur with analog video systems, but is not a problem for digital video cameras.

Three-dimensional recording procedures

Some, but not all, of the considerations of the previous section also apply to three-dimensional videography. The major requirements of three-dimensional analysis are discussed in this section. It is recommended that readers should gain good experience in all aspects of two-dimensional quantitative analysis before attempting three-dimensional analysis.

- At least two cameras are needed to reconstruct the three-dimensional movement-space coordinates of a point. The cameras should ideally be genlocked to provide shutter (or 'time') synchronisation; as we have noted several times above, this is not possible with most digital video cameras, which make no provision for such synchronisation.
- Synchronisation is also needed to link the events being recorded by the two cameras (event synchronisation). Event synchronisation can be achieved by the use of synchronised character generators if the cameras can be genlocked. For cameras that

cannot be genlocked, event and time synchronisation can be achieved by placing a timing device, such as a digital clock, in the fields of view of all cameras. Time synchronisation must then be performed mathematically at a later stage; obviously some error is involved in this process. It may also not be possible, particularly in competition, to include a timing device in the fields of view of the cameras. Some other event synchronisation must then be used, based on information available from the recorded sports movement, such as the instant of take-off in a jump.

- From two or more sets of image coordinates, some method is needed to reconstruct the three-dimensional movement-space coordinates. Several algorithms can be used for this purpose and the choice of the algorithm may have some procedural implications. Most of these algorithms involve the explicit or implicit reconstruction of the line (or ray) from each camera that is directed towards the point of interest, such as a skin marker. The location of that point is then estimated as that which is closest to the intersection of the rays from the two or more cameras.

- The simplest algorithm requires two cameras to be aligned with their optical axes perpendicular to each other. The cameras are then largely independent and the depth information from each camera is used to correct for perspective error for the other. The alignment of the cameras in this technique is difficult, although the reconstruction equations are very simple. This technique is generally too restrictive for use in sports competitions, where flexibility in camera placements is beneficial and sometimes essential.

- Flexible camera positions can be achieved with the most commonly used reconstruction algorithm, the 'direct linear transformation (DLT)'. This transforms the video image coordinates to movement-space coordinates by camera calibration involving independently treated transformation parameters for each camera. The algorithm requires a minimum of six calibration points with known three-dimensional coordinates and measured image coordinates to establish the DLT (transformation) parameters, or coefficients, for each camera independently. The DLT parameters incorporate the optical parameters of the camera and linear lens distortion factors. Because of the errors in sports biomechanical data, the DLT equations also incorporate residual error terms. The equations can then be solved directly by minimisation of the sum of the squares of the residuals. Once the DLT parameters have been established for each camera, the unknown movement-space coordinates of other points, such as skin markers, can then be reconstructed using the DLT parameters and the image coordinates for all cameras. Additional DLT parameters can also be included, if necessary, to allow for symmetrical lens distortion and asymmetrical lens distortions caused by decentring of the lens elements. No improvements in accuracy are usually achieved by incorporating non-linear lens distortions. The DLT algorithms impose several experimental restrictions.

  o An array of calibration (or control) points is needed, the coordinates of which are accurately known with respect to three mutually perpendicular axes. This is usually provided by some form of calibration frame (for example, Figures 4.4 and 4.7) or similar structure. The accuracy of the calibration coordinates is paramount, as it determines the maximum accuracy of other measurements. In filming activities such as kayaking, ski jumping, skiing, or javelin flight just after release, a group of vertical calibration poles, on which markers have been carefully positioned, may be easier to use than a calibration frame. The co-ordinates of the markers on the poles must be accurately measured; this often requires the use of surveying equipment. The use of calibration poles is generally more flexible and allows for a larger calibration volume than does a calibration frame.

o The more calibration points used, the stronger and more reliable is the reconstruction. It is often convenient to define the reference axes to coincide with directions of interest for the sports movement being investigated. The usual convention is for the x-axis to correspond with the main direction of horizontal motion.

o All the calibration points must be visible to each camera and their image coordinates must be clearly and unambiguously distinguishable, as in Figure 4.7. Calibration poles or limbs of calibration frames should not, therefore, overlap or nearly overlap, when viewed from any camera.

o Although an angle of 90° between the optical axes of the cameras might be considered ideal, deviations from this can be tolerated if kept within a range of about 60–120°. The cameras should also be placed so as to give the best views of the performer.

o Accurate coordinate reconstruction can only be guaranteed within the space – the calibration or control volume – defined by the calibration (or control)

Figure 4.7 Three-dimensional DLT camera set-up – note that the rays from the calibration spheres are unambiguous for both cameras – for clarity only the rays from all the upper or lower spheres are traced to one or other camera.

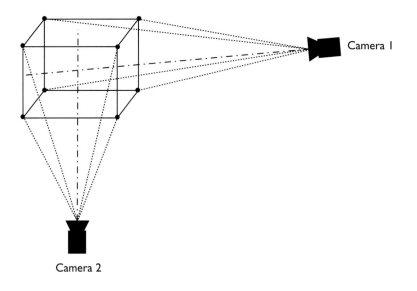

Camera 1

Camera 2

points. These should, therefore, be equally distributed within or around the volume in which the sports movement takes place. Errors depend on the distribution of the calibration points and increase if the performer moves beyond the confines of the control volume. This is the most serious restriction on the use of the DLT algorithm. It has led to the development of other methods that require smaller calibration objects, or the 'wand' technique now used by many automatic marker-tracking systems. All of these alternatives have greater computational complexity than the relatively straightforward DLT algorithm.

- If 'joint centre markers' are used, even greater attention must be paid to their movement relative to underlying bones than in a two-dimensional study. The markers, whether points or bands (see above) are only a guide to the location of the underlying joint centres of rotation. To minimise errors, you need an even more thorough anatomical knowledge of the joints and the location of their axes of rotation with respect to superficial landmarks throughout the range of segmental orientations than for a two-dimensional study. This becomes even more essential if markers are not used.

- Panning cameras can be used to circumvent the problem of a small image size, which would prevent identification of body landmarks if the control volume was very large. Three-dimensional reconstruction techniques that allow two cameras to rotate freely about their vertical axis (panning) and horizontal axis (tilting) have been developed. In these techniques, the cameras must be in known positions. These approaches have also been extended to allow for variation of the focal lengths of the camera lenses during filming.

- It is obviously necessary to check the validity of these methods. This can be done by calculating the root mean square (RMS) error between the reconstructed and known three-dimensional coordinates of points, preferably ones that have not been used to determine the DLT parameters. Furthermore, the success with which these methods reproduce three-dimensional movements can be checked, for example by filming the three-dimensional motion of a body segment of known dimensions or a rod thrown into the air. In addition, reliability and objectivity checks should be carried out on the digitised data.

DATA PROCESSING

The data obtained from digitising, either before or after transformation to three-dimensional coordinates, are often referred to as 'raw' data. Many difficulties arise when processing raw kinematic data and this can lead to large errors. As noted in the previous section, some errors can be minimised by careful equipment selection and rigorous attention to experimental procedures. However, the digitised coordinates will still contain random errors (noise).

The importance of this noise removal can be seen from consideration of an extremely simplified representation of a recorded sports movement, with a coordinate (*r*) expressed by the equation:

$$r = 2 \sin 4\pi t + 0.02 \sin 40\pi t$$

The first term on the right-hand side of this equation ($2 \sin 4\pi t$) represents the motion being observed (known, in this context, as the 'signal'). The amplitude of this signal is 2 – in arbitrary units – and its frequency is 4π radians/s (or 2 Hz) – indicated by $\sin 4\pi t$. The second term is the noise; this has an amplitude of only 1% of the signal (this would be a low value for many sports biomechanics studies) and a frequency of 40π rad/s (or 20 Hz), 10 times that of the signal. The difference in frequencies is because human movement generally has a low-frequency content and noise is at a higher frequency. Figure 4.8(a) shows the signal with the noise superimposed; note that there is little difference between the noise-free and noisy displacements. The above equation can now be differentiated to give velocity (*v*), which in turn can be differentiated to give acceleration (*a*). Then:

$$v = 8\pi \cos 4\pi t + 0.8\pi \cos 40\pi t$$

$$a = -32\pi^2 \sin 4\pi t - 32\pi^2 \sin 40\pi t$$

The noise amplitude in the velocity is now 10% ($0.8\pi/8\pi \times 100$) of the signal amplitude (Figure 4.8(b)). The noise in the acceleration data has the same amplitude, $32\pi^2$, as the signal, which is an intolerable error (Figure 4.8(c)). Unless the errors in the displacement data are reduced by smoothing or filtering, they will lead to considerable inaccuracies in velocities and accelerations and any other derived data. This will be compounded by any errors in body segment data (see pages 137–9).

Data smoothing, filtering and differentiation

Much attention has been paid to the problem of removal of noise from discretely sampled data in sports biomechanics. Solutions are not always (or entirely) satisfactory, particularly when transient signals, such as those caused by foot strike or other impacts, are present. Noise removal is normally performed after reconstruction of the movement coordinates from the image coordinates because, for three-dimensional studies, each set of image coordinates does not contain full information about the movement co-ordinates. However, the noise removal should be performed before calculating other data, such as segment orientations and joint forces and moments. The reason for this is that the calculations are highly non-linear, leading to non-linear combinations of random noise, which can adversely affect the separation of signal and noise by low-pass filtering.

The three most commonly used techniques to remove high-frequency noise from the low-frequency movement coordinates use digital low-pass filters, usually Butterworth filters or Fourier series truncation, or spline smoothing; the last of these is normally

Figure 4.8 Simple example of noise-free data, shown by dashed black lines, and noisy data, shown by blue lines: (a) displacement; (b) velocity; (c) acceleration (all in arbitrary units).

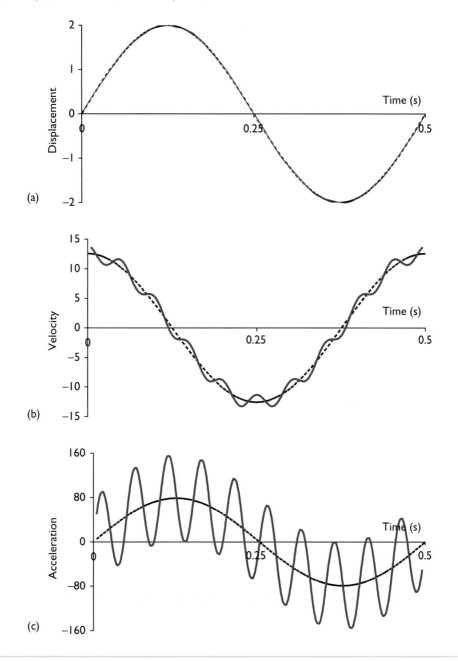

realised through cross-validated quintic splines. The first or last of these are used in most commercial quantitative analysis packages. Details of these three techniques are included in Appendix 4.1 for interested readers.

- Quintic splines appear to produce more accurate first and second derivatives than most other techniques that are commonly used in sports biomechanics. The two filtering techniques (Fourier truncation and digital filters) were devised for periodic data, where the pattern of movement is cyclical, as in Figure 4.8(a). Sporting activities that are cyclic, such as running, are obviously periodic, and some others can be considered quasi-periodic. Problems may be encountered in trying to filter non-periodic data, although these may be overcome by removing any linear trend in the data before filtering; this makes the first and last data values zero. The Butterworth filter often creates fewer problems here than Fourier truncation, but neither technique deals completely satisfactorily with constant acceleration motion, as for the centre of mass when a sports performer is airborne.
- The main consideration for the sports biomechanist using smoothing or filtering routines is a rational choice of filter cut-off frequency or spline smoothing parameter. A poor choice can result in some noise being retained if the filter cut-off frequency is too high, or some of the signal being rejected if the cut-off frequency is too low. As most human movement is at a low frequency, a cut-off frequency of between 4 and 8 Hz is often used. Lower cut-off frequencies may be preferable for slow events such as swimming, and higher ones for impacts or other rapid energy transfers. The cut-off frequency should be chosen to include the highest frequency of interest in the movement. As filters are sometimes implemented as the ratio of the cut-off to the sampling frequency in commercially available software, an appropriate choice of the latter might need to have been made at an earlier stage.
- The need for data smoothness demands a minimum ratio of the sampling to cut-off frequencies of 4:1, and preferably one as high as 8:1 or 10:1. The frame rate used when video recording, and the digitising rate (the sampling rate), must allow for these considerations.
- The use of previously published filter cut-off frequencies or manual adjustment of the smoothing parameter is not recommended. Instead, a technique should be used that involves a justifiable procedure to take into account the peculiarities of each new set of data. Attempts to base the choice of cut-off frequency on some objective criterion have not always been successful. One approach is to compare the RMS difference between the noisy data and that obtained after filtering at several different cut-off frequencies with the standard deviation obtained from repetitive digitisation of the same anatomical point. The cut-off frequency should then be chosen so that the magnitudes of the two are similar. Another approach is called residual analysis, in which the residuals between the raw and filtered data are calculated for a range of cut-off frequencies: the residuals are then plotted against the cut-off frequency, and the best value of the latter is chosen as that at which the residuals begin to approach an asymptotic value, as in Figure 4.9; some subjective judgement is involved in assigning the cut-off frequency at which this happens.

Figure 4.9 Residual analysis of filtered data.

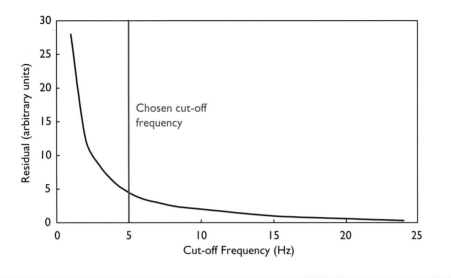

- It is often necessary to use a different smoothing parameter or cut-off frequency for the coordinates of the different points recorded. This is particularly necessary when the frequency spectra for the various points are different.
- Finally, you should note that no automatic noise-removal algorithm will always be successful, and that the smoothness of the processed data should always be checked.

Body segment inertia parameters

Various body segment inertia parameters are used in movement analysis. The mass of each body segment and the segment centre of mass position are used in calculating the position of the whole body centre of mass (see Chapter 5). These values, and segment moments of inertia, are used in calculations of net joint forces and moments using the method of inverse dynamics. The most accurate and valid values available for these inertia parameters should obviously be used. Ideally, they should be obtained from, or scaled to, the sports performer being studied. The values of body segment parameters used in sports biomechanics have been obtained from cadavers and from living persons, including measurements of the performers being filmed.

Cadaver studies have provided very accurate segmental data. However, limited sample sizes throw doubt on the extrapolation of these data to a general sports population. They are also highly questionable because of the unrepresentative samples in respect of sex, age and morphology. Problems also arise from the use of different dissection techniques by different researchers, losses in tissue and body fluid during dissection and degeneration associated with the state of health preceding death. Segment mass may be expressed as a simple fraction of total body mass or, more accurately, in the form of a linear regression equation with one or more anthropometric variables. Even the latter may cause under- or over-estimation errors of total body mass as large as 4.6%.

Body segment data have been obtained from living people using gamma-ray scanning or from imaging techniques, such as computerised tomography and magnetic resonance imaging. These may eventually supersede the cadaver data that are still too often used.

Obtaining body segment parameter data from sports performers may require sophisticated equipment and a great deal of the performer's time. The immersion technique is simple, and can be easily demonstrated by any reader with a bucket, a vessel to catch the overflow from the bucket, and some calibrated measuring jugs or a weighing device (see Figure 4.10 and Study task 5). It provides accurate measurements of segment volume and centre of volume, but requires a knowledge of segmental density to calculate segment mass. Also, as segment density is not uniform throughout the segment, the centre of mass does not coincide with the centre of volume.

Figure 4.10 Simple measurement of segment volume.

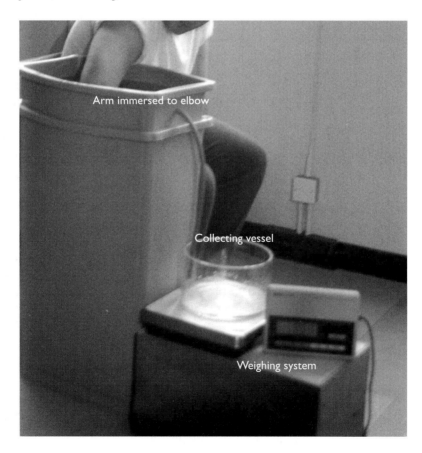

There are several 'mathematical' models that calculate body segment parameters from standard anthropometric measurements, such as segment lengths and circumferences. Some of these models result in large errors even in estimates of segment volumes. Others are very time-consuming, requiring up to 200 anthropometric measurements, which take at least an hour or two to complete. All these models require density values from other sources, usually cadavers, and most of them assume constant density throughout the segment, or throughout large parts of the segment.

The greatest problems in body segment data occur for moments of inertia. There are no simple yet accurate methods of measuring segmental moments of inertia for a living person. Many model estimations are either very inaccurate or require further validation. A relative error of 5% in segmental moments of inertia may be quite common. Norms or linear regression equations are often used, but these should be treated with caution as the errors involved in their use are rarely fully assessed. It may be necessary to allow for the non-linear relationships between segmental dimensions and moment of inertia values.

Data errors

Uncertainties, also referred to as errors, in the results of biomechanical data processing can be large, particularly for computation of kinetic variables. This is mainly because of errors in the body segment data and linear and angular velocities and accelerations, and the combination of these errors in the inverse dynamics equations. If such computations are to be attempted, scrupulous adherence to good experimental protocols is essential. A rigorous assessment of the processing techniques is also necessary. The topic of error analysis is a very important one and the value of sports biomechanical measurements cannot be assessed fully in the absence of a quantification of the measurement error. The accuracy of the measuring system and the precision of the measurements should be assessed separately. Error propagation in calculations can be estimated using standard formulae (see Challis, 2007; Further Reading, page 152).

PROJECTILE MOTION

In this section, we illustrate an important example of linear (in fact, curvilinear) motion – the motion of a projectile in the air. Projectiles are bodies launched into the air that are subject only to the forces of gravity and air resistance. Projectile motion occurs frequently in sport and exercise activities. Often the projectile involved is an inanimate object, such as a shot or golf ball. In some activities the sports performer becomes the projectile, as in the long jump, high jump, diving and gymnastics. An understanding of the mechanical factors that govern the flight path or trajectory of a projectile is, therefore, important in sports biomechanics. The following discussion assumes that the effects of aerodynamic forces – both air resistance and more complex lift effects – on

BOX 4.3 THOSE THINGS CALLED VECTORS AND SCALARS

Vector algebra has been the bane of sports biomechanics for many students who want to work in the real world of sport and exercise, providing scientific support rather than doing research. Even many quantitative biomechanists do not use vector algebra routinely in their work. This box introduces the basics, which all sports biomechanics students should know. Appendix 4.2 includes some further vector algebra for interested students.

We distinguish between scalar variables, which have a magnitude but no directional quality, and vectors. Vectors have both a magnitude and a direction; their behaviour cannot be studied only by their magnitude – the direction is also important. This means that they cannot be added, subtracted or multiplied as scalars, for which we use simple algebra and arithmetic.

Mass, volume, temperature and energy are scalar quantities. Some of these are always positive – such as mass, volume and kinetic energy – whereas others depend upon arbitrary choices of datum and can be both positive and negative – temperature and potential energy come into this category. A third group often use a convention to designate a 'direction' in which the scalar 'moves'; for example, the work done by a biomechanical system on its surroundings – as in a muscle raising a weight is, by convention, considered positive, whereas when the surroundings do work on the system, this is considered to be negative. Work remains, however, a scalar.

Many of the kinematic variables that are important for the biomechanical understanding and evaluation of movement in sport and exercise are vector quantities. These include linear and rotational position, displacement, velocity and acceleration. Many of the kinetic variables considered in Chapter 5, such as momentum, force and torque, are also vectors. The magnitude – the scalar part of the vector if you like – of a kinematic variable usually has a name that is different from that of the vector (see Table 4.1). Interestingly, because a scalar has no implied direction, speed should always be positive: designating it as positive up or to the left and negative down or to the right converts it into a one-dimensional vector. Note that the SI system does not recognise degrees/s as a preferred unit, but I have included it in Table 4.1 as it means more to most students (and to me) than the approved unit, radians/s (π radians = 180°).

Table 4.1 Kinematic vectors and scalars

VECTOR AND SYMBOL	SCALAR AND SYMBOL	SI UNIT
Linear		
Displacement, s or $\mathbf{s}$	Distance, s	metres, m
Velocity, v or $\mathbf{v}$	Speed, v	metres per second, m/s
Acceleration, a or $\mathbf{a}$	Acceleration, a	metres per second per second, m/s^2
Angular		
Angular displacement, $\boldsymbol{\theta}$	Angular distance, θ	radians, rad; degrees, °
Angular velocity, $\boldsymbol{\omega}$	Angular speed, ω	radians per second, rad/s
		degrees per second, °/s
Angular acceleration, $\boldsymbol{\alpha}$	Angular acceleration, α	radians per second per second, rad/s^2
		degrees per second per second, °/s^2

Scalar variables are shown in italicised type, except for Greek symbols which are not italicised, as above. Vector quantities are shown in bold type, sometimes italicised, as above. Linear vectors can be represented graphically by a straight line arrow in the direction of the vector, with the length of the line being proportional to the magnitude of the vector. Angular vectors can also be represented graphically; in this case the direction of the arrow is found from the right-hand rule, shown diagrammatically in Figure 4.11. With the right hand orientated as in Figure 4.11, the curled fingers follow the direction of rotation and the thumb points in the direction of the angular motion vector. So, for example, the direction of the angular motion vectors (angular displacement, velocity and acceleration) for flexion–extension of the knee joint, a movement in the sagittal plane, lies along the flexion–extension axis, the transverse axis perpendicular to the sagittal plane. In the case shown here, the angular motion vector and the axis of rotation coincide. This is often, but not always, the case. See Appendix 4.2 for further information about vectors.

Figure 4.11 The right-hand rule.

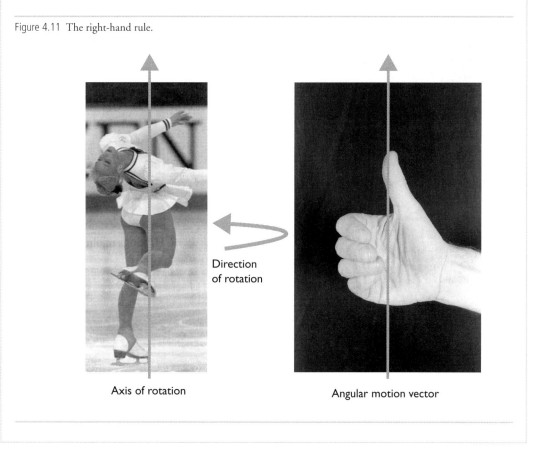

Direction of rotation

Axis of rotation

Angular motion vector

projectile motion are negligible. This is a reasonable first assumption for some, but certainly not all, projectile motions in sport; aerodynamic forces will be covered in Chapter 5. Although I introduce algebraic symbols and equations in what follows, for shorthand, I do not want to focus on the mathematical derivation of these equations, only on their importance.

There are three parameters, in addition to gravitational acceleration, g, that determine the trajectory of a simple projectile, such as a ball, shot or hammer. These are the projection speed, angle and height (Figure 4.12). For thrown objects, these three parameters are often called release speed, angle and height; for humans, the terms take-off speed, angle and height are more common. I use projection speed, angle and height to cover both types of 'projectile', objects and humans, and cover these three parameters in decreasing order of importance.

Projection speed

Projection speed (v_0) is defined as the speed of the projectile at the instant of release or take-off (Figure 4.12). When the projection angle and height are held constant, the projection speed will determine the maximum height the projectile reaches (its apex) and its range, the horizontal distance it travels. The greater the projection speed, the greater the apex and range. It is common practice to resolve a projectile's velocity vector into its horizontal and vertical components and then to analyse these independently. Horizontally a projectile is not subject to any external forces, as we are ignoring air resistance, and will therefore have a constant horizontal velocity while in the air, as in a long jump or swimming start dive. The range (R) travelled by a projectile is the product of its horizontal projection velocity ($v_{x0} = v \cos\theta$) and its time of flight (t_{max}). That is:

$$R = v_{x0}\, t_{max}$$

To calculate a projectile's time of flight, we must consider the magnitude of the vertical component of its projection velocity (v_y). Vertically a projectile is subject to a constant acceleration due to gravity (g). The magnitude of the maximum vertical displacement (y_{max}), flight time (t_{max}) and range (R) achieved by a projectile can easily be determined from v_{y0} if it takes off and lands at the same level ($y_0 = 0$). This occurs, for example, in a football kick. In this case, the results are as follows:

$$y_{max} = v_{y0}^2/2g = v_0^2\sin^2\theta/2g$$
$$t_{max} = 2v_{y0}/g = 2v_0 \sin\theta/g$$
$$R = 2v_0^2\sin\theta \cos\theta/g = v_0^2\sin2\theta/g$$

The range (R) equation shows that the projection speed is by far the most important of the projection parameters in determining the range achieved, because the range is proportional to the square of the release speed. Doubling the release speed would increase the range four-fold.

Figure 4.12 Projection parameters.

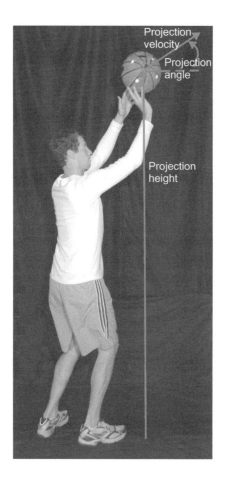

Projection angle

The projection angle (θ) is defined as the angle between the projectile's line of travel (its velocity vector) and the horizontal at the instant of release or take-off. The value of the projection angle depends on the purpose of the activity. For example, activities requiring maximum horizontal range, such as the shot put, long jump and ski jump, tend to use smaller angles than those in which maximum height is an objective, for example the high jump or the jump of a volleyball spiker. In the absence of aerodynamic forces, all projectiles will follow a flight path with a parabolic shape that depends upon the projection angle (Figure 4.13).

Figure 4.13 Effect of projection angle on shape of parabolic trajectory for a projection speed of 15 m/s and zero projection height. Projection angles: 30° – continuous black curve, flight time 1.53 s, range 19.86 m, maximum height 2.87 m; 45° – blue curve, flight time 2.16 s, range 22.94 m, maximum height 5.73 m; 60° – dashed black curve, flight time 2.65 s, range 19.86 m, maximum height 8.60 m.

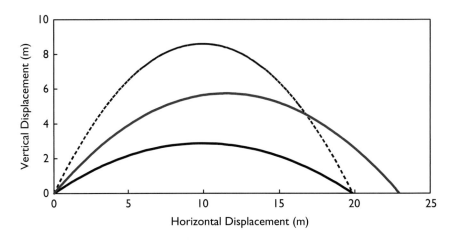

Projection height

The equations relating to projection speed (page 142) have to be modified if the projectile lands at a height higher or lower than that at which it was released. This is the case with most sports projectiles, for example in a shot put, a basketball shot or a long jump. For a given projection speed and angle, the greater the projection height (y_0), the longer the flight time and the greater the range and maximum height. The maximum height is the same as above but with the height of release added, as follows:

$$y_{max} = y_0 + v_0^2 \sin^2\theta/2g$$

$$t_{max} = v_0 \sin\theta/g + (v_0^2\sin^2\theta + 2gy_0)^{1/2}/g$$

$$R = v_0^2\sin2\theta/2g + v_0^2\cos\theta(\sin^2\theta + 2gy_0/v_0^2)^{1/2}/g$$

The equations for the time of flight (t_{max}) and range (R) appear to be much more complicated than those for zero release height. The first of the two terms in each of these relates, respectively, to the time and horizontal distance to the apex of the trajectory. The value of these terms is exactly half of the total values in the earlier equations. The second terms relate to the time and horizontal distance covered from the apex to landing. By setting the release height (y_0) to zero, and noting that $\cos\theta\sin\theta = \cos2\theta/2$, you will find that the second terms in the equations for the time of flight and range become equal to the first terms and that the equations are then identical to the earlier ones for which the release height was zero.

Optimum projection conditions

In many sports events, the objective is to maximise either the range, or the height of the apex achieved by the projectile. As seen above, any increase in projection speed or height is always accompanied by an increase in the range and height achieved by a projectile. If the objective of the sport is to maximise height or range, it is important to ascertain the best – the optimum – angle to achieve this. Obviously maximum height is achieved when all of the available projection speed is directed vertically, when the projection angle is 90°.

As we saw above, when the projection height is zero, the range is given by $v_0^2 \sin 2\theta / g$. For a given projection speed, v_0, the range is a maximum when $\sin 2\theta$ is a maximum, that is when $\sin 2\theta = 1$ and $2\theta = 90°$; therefore, the optimum projection angle, θ, is 45°. For the more general case of a non-zero projection height, the optimum projection angle can be found from: $\cos 2\theta = g y_0 / (v_0^2 + g y_0)$. For a good shot putter, for example, this would give a value of around 42°.

Although optimum projection angles for given values of projection speed and height can easily be determined from the last equation, they do not always correspond to those recorded from the best performers in sporting events. This is even true for the shot put, for which the object's flight is the closest to a parabola of all sports objects. The reason for this is that the calculation of an optimum projection angle assumes, implicitly, that the projection speed and projection angle are independent of one another. For a shot putter, the release speed and angle are, however, not independent, because of the arrangement and mechanics of the muscles used to generate the release speed of the shot. A greater release speed, and hence range, can be achieved at an angle of about 35°, which is less than the optimum projectile angle. If the shot putter seeks to increase the release angle to a value closer to the optimum projectile angle, the release speed decreases and so does the range.

A similar deviation from the optimum projection angle is noticed when the activity involves the projection of an athlete's body. The angle at which the body is projected at take-off can have a large effect on the take-off speed. In the long jump, for example, take-off angles used by elite long jumpers are around 20°. To obtain the theoretically optimum take-off angle of around 42°, long jumpers would have to decrease their normal horizontal speed by around 50%. This would clearly result in a drastically reduced range, because the range depends largely on the square of the take-off speed.

In many sporting events, such as the javelin and discus throws, badminton, sky-diving and ski jumping, the aerodynamic characteristics of the projectile can significantly influence its trajectory. The projectile may travel a greater or lesser distance than it would have done if projected in a vacuum. Under such circumstances, the calculations of optimal projection parameters need to be modified considerably to take account of the aerodynamic forces (see Chapter 5) acting on the projectile.

LINEAR VELOCITIES AND ACCELERATIONS CAUSED BY ROTATION

In this section, we show how the rotation of a body can cause linear motion. Consider the quasi-rigid body of a gymnast shown in Figure 4.14, rotating about an axis fixed in the bar, with angular velocity ω, and angular acceleration, α. The axis of rotation in the bar is fixed and we will assume that the position of the gymnast's centre of mass is fixed relative to that axis. Therefore, the distance of the centre of mass from the axis of rotation (the magnitude r of the position vector r) does not change.

The linear velocity of the centre of mass of the gymnast is tangential to the circle that the centre of mass describes; its magnitude is given by $v = \omega\, r$. Readers who are familiar with basic vector algebra, or those of you who can follow Appendix 4.2, will appreciate that the tangential velocity of the centre of mass, from the cross-product rule (Appendix 4.2), is mutually perpendicular to both the position (r) and angular velocity (ω) vectors. The angular velocity vector has a direction, perpendicularly out of the plane of the page towards you, determined by the right-hand rule of Figure 4.11.

The linear acceleration of the centre of mass of the gymnast has two components, which are mutually perpendicular, as follows.

- The first component, called the tangential acceleration, has a direction identical to that of the tangential velocity and a magnitude $\alpha\, r$.
- The second component is called the centripetal acceleration. It has a magnitude $\omega^2 r$ ($= \omega\, v = v^2/r$) and a direction given by the vector cross-product (Appendix 4.2) $\omega \times v$. This direction is obtained by letting the angular velocity vector (ω) rotate towards the tangential velocity vector (v) through the right angle between them and using the vector cross-product rule of Appendix 4.2. This gives a vector ($\omega \times v$) perpendicular to both the angular velocity and tangential velocity vectors, as shown in Figure 4.14. As the velocity vector is perpendicular to the longitudinal axis of the gymnast, the direction of the centripetal acceleration vector ($\omega \times v$) is along the longitudinal axis of the gymnast towards the axis of rotation. You should note that a centripetal acceleration exists for all rotational motion. This is true even if the angular acceleration is zero, in which case the angular velocity is constant; the magnitude of the centripetal acceleration is also constant if the distance from the axis of rotation to the centre of mass is constant.

ROTATION IN THREE-DIMENSIONAL SPACE

For two-dimensional rotations of human body segments, the joint angle may be defined as the angle between two lines representing the proximal and distal segments. A similar procedure can be used to specify the relative angle between line representations of segments in three dimensions if the articulation is a simple hinge joint, but generally the process is more complex. There are many different ways of defining the orientation angles of two articulating rigid bodies and of specifying the orientation angles of the

Figure 4.14 Tangential velocity and tangential and centripetal acceleration components for a gymnast rotating about a bar. Her angular velocity vector lies along the bar as shown by the blue arrow; the position vector (dashed white arrow) runs from the bar to the gymnast's centre of mass. The tangential velocity vector (black arrow) is perpendicular to the position vector and passes posterior to her centre of mass. The tangential acceleration vector is also represented by the black arrow. The centripetal acceleration vector, shown by the continuous white arrow, is in the direction opposite to that of the position vector.

human performer. Most of these conventions have certain problems, one of which is to have an angle convention that is easily understood. Further discussion of this topic is beyond the scope of this book (but see Milner, 2007; Further Reading, page 152).

The specification of the angular orientation of the human performer as a whole is also problematic. The representation of Figure 4.15, for example, has been used to analyse airborne movements in gymnastics, diving and trampolining. In this, rotation is specified by the somersault angle (φ) about a horizontal axis through the centre of mass, the twist angle (ψ) about the longitudinal axis of the performer, and the tilt angle (θ). The last named is the angle between the longitudinal axis and the fixed plane normal to the somersault angular velocity vector.

Figure 4.15 Angular orientation showing angles of somersault (φ), tilt (θ) and twist (ψ).

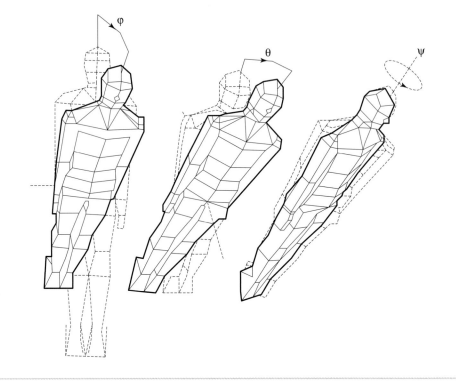

SUMMARY

In this chapter, we covered the use of videography in the study of sports movements, including the equipment and methods used. The necessary features of video equipment for recording movements in sport were considered, as were the advantages and limitations of two- and three-dimensional recording of sports movements. We outlined the possible sources of error in recorded movement data and described experimental procedures that will minimise recorded errors in two- and three-dimensional movements. The need for smoothing or filtering of kinematic data was covered and the ways of performing this were touched on. We also outlined briefly the requirement for accurate body segment inertia parameter data and how these can be obtained, and some aspects of error analysis. Projectile motion was considered and equations presented to calculate the maximum vertical displacement, flight time, range and optimum projection angle of a simple projectile for specified values of the three projection parameters. Deviations of the optimal angle for the sports performer from the optimal projection angle were explained. We looked at the calculation of linear velocities and accelerations caused by rotation and concluded with a brief consideration of three-dimensional rotation.

STUDY TASKS

1 Explain why quantitative video analysis is important in the study of sports techniques.

Hint: You may find it useful to reread the section on 'Comparison of qualitative and quantitative movement analysis' in Chapter 1 (pages 36–40) as well as the section on 'The use of videography in recording sports movements' near the start of this chapter (pages 117–20).

2 Obtain a video recording of a sports movement of your choice. Study the recording carefully, frame by frame. Identify and describe important aspects of the technique, such as key events that separate the various phases of the movement. Also, identify the displacements, angles, velocities and accelerations that you would need to include in a quantitative analysis of this technique. If you have access to a video camera, you may wish to choose the sports movement you will use in Study task 6.

Hint: You may find it useful to reread the section on 'Identifying critical features of a movement' in Chapter 2 (pages 59–72) before undertaking this task.

3 (a) List the possible sources of error in recorded movement data, and identify which would lead to random and which to systematic errors.

(b) Briefly describe the procedures that would minimise the recorded error in a study of an essentially two-dimensional movement. Assess which of these steps could not be implemented if video recording at an elite sports competition. Briefly explain how these procedures would be modified for recording a three-dimensional movement.

Hint: You should reread the section on 'Experimental procedures' (pages 126–33) before undertaking this task.

4 Download the Excel workbook containing the data from one of the five speeds for the walk-to-run transition study on the book's website. The knee angles contained in that workbook were calculated from filtered linear coordinates that had been auto-tracked using markers. We will assume therefore that the angles in the workbook are sufficiently noise-free to be able to estimate accurately the knee angular velocities and accelerations from the simple numerical differentiation equations below.

(a) Calculate the knee angular velocities (ω_i) as follows, using Excel:

$\omega_i = (\theta_{i+1} - \theta_{i-1})/(2\Delta t)$, where ω_i is the angular velocity at time interval (Excel row) number i and θ_{i+1} and θ_{i-1} are the knee angles at time intervals (rows) i+1 and i−1, respectively. The denominator in the equation ($2\Delta t$) is the time interval between times (rows) i+1 and i−1 and is 0.0392 s. Tabulate your knee angular velocity data in a new column in your Excel worksheet. Note that we cannot estimate the angular velocities at the first and last instants of the knee angle–time series.

(b) Calculate the knee angular accelerations (α_i) as follows, using Excel:

$\alpha_i = (\omega_{i+1} - \omega_{i-1})/(2\Delta t)$, where α_i is the angular acceleration at time interval (Excel row) number i and ω_{i+1} and ω_{i-1} are the knee angular velocities at time intervals (rows) i+1 and i−1, respectively. As before, the denominator in the equation (2Δt) is the time interval between times (rows) i+1 and i−1 and is 0.0392 s. Tabulate your knee angular acceleration data in a new column in your Excel worksheet. Note that we cannot estimate the angular accelerations at the first and last instants of the knee angular velocity–time series (and, therefore, not at the first two and last two time instants of the knee angle–time series).

(c) Plot the time-series graphs of knee angle, angular velocity and angular acceleration. Do the time-series patterns basically agree with the qualitative patterns for running from Figure 3.10? Explain your answer. Does the angular acceleration–time series graph look sufficiently 'smooth' to justify our assumption about the knee angle data being noise-free? Comment on your findings.

Hint: You may wish to reread the section in Chapter 3 on 'The geometry of angular motion' (pages 93–6) and consult your answer to Study task 4 in that chapter before undertaking this task.

5 Carry out an experiment to determine the volume of (a) a hand, (b) a forearm segment. You will need a bucket or similar vessel large enough for the hand and forearm to be fully submerged. You will also need a bowl, or similar vessel, in which the bucket can be placed, to catch the overflow of water; and calibrated containers to measure the volume of water. Repeat the experiment at least three times and then calculate the mean volume and standard deviation for each segment. How reproducible are your data?

Hint: You may wish to reread the subsection on 'Body segment inertia parameters' (pages 137–9) before undertaking this task.

6 (a) Plan an experimental session in which you would record an essentially two-dimensional sports movement, such as a long jump, running or a simple gymnastics vault. You should carefully detail all the important procedural steps, including the use of skin markers (see Table 5.1 and Box 6.2).

(b) If you have access to a suitable video camera, record several trials of the movement from one or more performers; if not, download a suitable running sequence from the book's website.

(c) If you have access to a video digitising system, then digitise at least one of the sequences you have recorded. If you have access to analysis software, then plot stick figure sequences and graphs of relevant kinematic variables, which should have been established by a qualitative analysis of the movement similar to that performed in Study task 2. If you do not have this access, you can download stick figure sequences and graphs of kinematic variables from the book's website, but you will still need to justify which variables are of interest.

(d) Write a short technical report of your study in no more than 1500 words. Focus on the important results.

Hint: You should reread the subsection on 'Two-dimensional recording procedures' (pages 126–30) before undertaking this task.

7 (a) A shot is released at a height of 1.89 m, with a speed of 13 m/s and at an angle of 34°. Calculate the maximum height reached and the time at which this

occurs, the range and the time of flight, and the optimum projection angle. Why do you think the release angle differs from the optimum projection angle?

(b) Calculate the maximum height reached and the time at which this occurs, the range and the time of flight, and the optimum projection angle for a similar object projected with zero release height but with the same release speed and angle. Comment on the effects of changing the release height.

Hint: You should reread the section on 'Projectile motion' (pages 139, 141–5) before undertaking this study task.

8 Calculate the magnitudes of the acceleration components and the velocity of a point on a rigid body at a radius of 0.8 m from the axis of rotation when the angular velocity and angular acceleration have magnitudes, respectively, of 10 rad/s and −5 rad/s^2. Sketch the body and draw on it the velocity vector and components of the acceleration vector.

Hint: You should reread the section on 'Linear velocities and accelerations caused by rotation' (page 146) before undertaking this task.

You should also answer the multiple choice questions for Chapter 4 on the book's website.

GLOSSARY OF IMPORTANT TERMS (compiled by Dr Melanie Bussey)

Analog Any signal continuous in both time and amplitude. It differs from a digital signal in that small fluctuations in the signal are meaningful. The primary disadvantage of analog signal processing is that any system has noise – random variation. As the signal is copied and recopied, or transmitted over long distances, these random variations become dominant. See also **digital**.

Calibration Refers to the process of determining the relationship between the output (or response) of a measuring instrument and the value of the input quantity or attribute used as a measurement standard.

Cinematography Motion picture photography. See also **videography**.

Cross-sectional (study) The observation of a defined population at an instant in time or across a specified time interval; exposure and outcome are determined simultaneously. See also **longitudinal study**.

Digital Uses discrete values (often electrical voltages), rather than a continuous spectrum of values as in an analog system, particularly those representable as binary numbers, or non-numeric symbols such as letters or icons, for input, processing, transmission, storage or display. A digital transmission (as for digital radio or television) is considered less 'noisy' because slight variations do not matter as they are ignored when the signal is received. See also **analog**.

Dimension A term denoting the spatial extent of a measurable quantity. See also **two-dimensional** and **three-dimensional**.

Digitising The process of specifying or measuring the x- and y-image coordinates of points on a video frame; more strictly called coordinate digitising.

Genlock A common technique in which the video output of one source, or a specific reference signal, is used to synchronise other picture sources so that images are captured simultaneously. Rarely found, currently, on digital video cameras.

Inverse dynamics An analytical approach calculating forces and moments based on the accelerations of the object, usually computed from measured displacements and angular orientations from videography or another image-based motion analysis system.

Longitudinal (study) A correlational research study that involves observations of the same items over long time periods. Unlike a **cross-sectional study**, a longitudinal study tracks the same people and, therefore, the differences observed in those people are less likely to be the result of cultural differences across generations.

Low-pass filter A filter that passes low frequencies but attenuates (or reduces) frequencies above the cut-off frequency. In movement analysis, used mainly to remove high-frequency 'noise' from a low-frequency movement signal.

Statics The branch of mechanics in which the system being studied undergoes no acceleration.

Three-dimensional Occurring in two or three planes; requiring a minimum of three coordinates to describe, for example x-, y- and z-coordinates. See also **two-dimensional**.

Two-dimensional Occurring within a single plane; requiring a minimum of two coordinates to describe, for example x- and y-coordinates. See also **three-dimensional**.

Videography The process of capturing images on a videotape or directly to a computer; also used to include the later analysis of these images. See also **cinematography**.

FURTHER READING

Bartlett, R.M. (1999) *Sports Biomechanics: Reducing Injury and Improving Performance*, London: E & FN Spon. Chapter 4 provides a simple introduction to inverse dynamics without being too mathematical.

Challis, J.H. (2007) Data processing and error estimation, in C.J. Payton and R.M. Bartlett (eds) *Biomechanical Evaluation of Movement in Sport and Exercise*, Abingdon: Routledge. Chapter 8 elaborates on the errors in sports biomechanical data and outlines, with clear examples, the calculation of uncertainties in derived biomechanical data.

Milner, C. (2007) Motion analysis using on-line systems, in C.J. Payton and R.M. Bartlett (eds) *Biomechanical Evaluation of Movement in Sport and Exercise*, Abingdon: Routledge. Chapter 3 provides a lucid and easy-to-follow explanation of some difficult concepts. Highly recommended if you see research into sports movement as something you might wish to pursue.

Payton, C.J. (2007) Motion analysis using video, in C.J. Payton and R.M. Bartlett (eds) *Biomechanical Evaluation of Movement in Sport and Exercise*, Abingdon: Routledge. Chapter 2 contains much useful advice on a video study, including the reporting of such a study, which you could adopt for the technical report in Study task 6.

APPENDIX 4.1 DATA SMOOTHING AND FILTERING

Digital low-pass filters

Digital low-pass filters are widely used to remove, or filter, high-frequency noise from digital data. Butterworth filters (of order $2n$ where n is a positive integer) are often used in sports biomechanics, because they have a flat passband, the band of frequencies that is not affected by the filter (Figure 4.16). However, they have relatively shallow cut-offs. This can be improved by using higher-order filters, but round-off errors in computer calculations can then become a problem. They also introduce a phase shift, which must be removed by a second, reverse filtering, which increases the order of the filter and further reduces the cut-off frequency.

Butterworth filters are recursive; that is, they use filtered values of previous data points as well as noisy data values to obtain filtered data values. This makes for faster computation but introduces problems at the ends of data sequences, at which filtered values must be estimated. This can mean that extra frames must be digitised at each end of the sequence and included in the data processing; these extra frames then have to be discarded after filtering. This can involve unwelcome extra work for the movement analyst; other solutions include various ways of padding the ends of the data sets.

The main decision for the user, as with Fourier series truncation, is the choice of cut-off frequency (discussed on pages 133–7). The filtered data are not obtained in analytic form, so a separate numerical differentiation process must be used. Although Butterworth filtering appears to be very different from spline fitting (see below), the two are, in fact, closely linked.

Figure 4.16 Low-pass filter frequency characteristics.

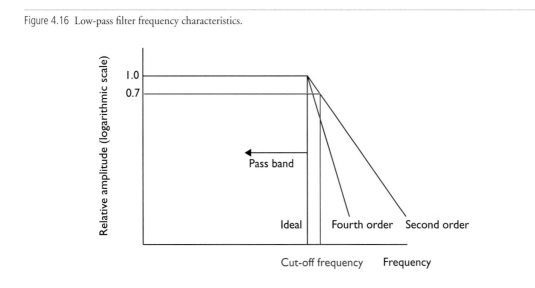

Fourier series truncation

The first step in Fourier series truncation is the transformation of the noisy data into the frequency domain by means of a Fourier transformation. This, in essence, replaces the familiar representation of displacement as a function of time (the time domain) as in Figure 4.17(a) by a series of sinusoidal waves of different frequencies. This 'frequency domain' representation of the data is then presented as amplitudes of the sinusoidal components at each frequency – the harmonic frequencies, as in Figure 4.17(b), or as a continuous curve. Figure 4.18(a) shows the frequency domain representation of the simplified data of Figure 4.17(a). The data are then filtered to remove high-frequency noise. This is done by reconstituting the data up to the chosen cut-off frequency and truncating the number of terms in the series from which it is made up. For the

Figure 4.17 Displacement data represented in: (a) the time domain; (b) the frequency domain.

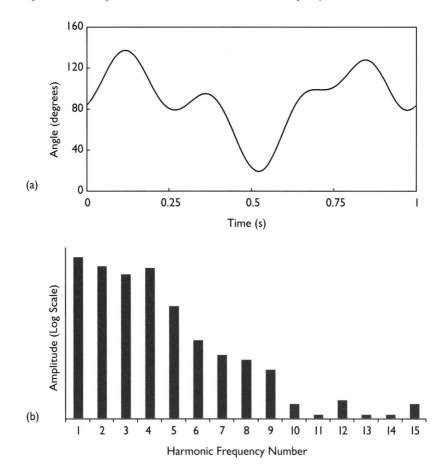

Figure 4.18 Simple example of: (a) noisy data in the frequency domain; (b) same data in the time domain after removal of noise by filtering by Fourier series truncation.

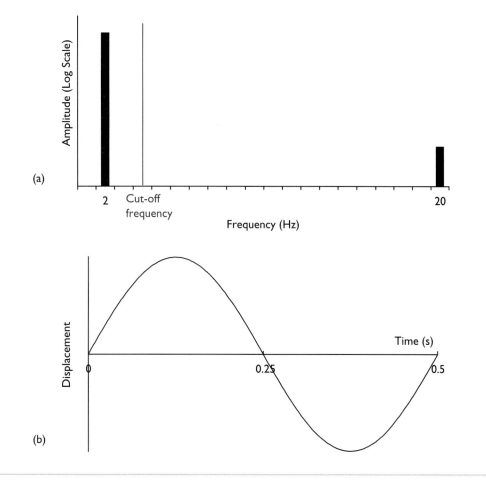

simplified data of Figure 4.18(a), if the cut-off frequency was, say, 4 Hz, then the second frequency term would be rejected as noise.

In this simplified case, the noise would have been removed perfectly, as the time domain signal of Figure 4.18(b) demonstrates. There would be no resulting errors in the velocities and accelerations. The major decision here concerns the choice of cut-off frequency, and similar principles to those described on pages 133–7 can be applied. Unlike digital filters, the cut-off can be infinitely steep, as in Figure 4.16, but this is not necessarily the case. The filtered data can be represented as an equation and can be differentiated analytically. This technique requires the raw data points to be sampled at equal time intervals, as do digital low-pass filters (see above).

Figure 4.19 Over-smoothing: (a) velocity; (b) acceleration. Under-smoothing: (c) velocity; (d) acceleration. The optimum smoothing is shown in (e) velocity and (f) acceleration.

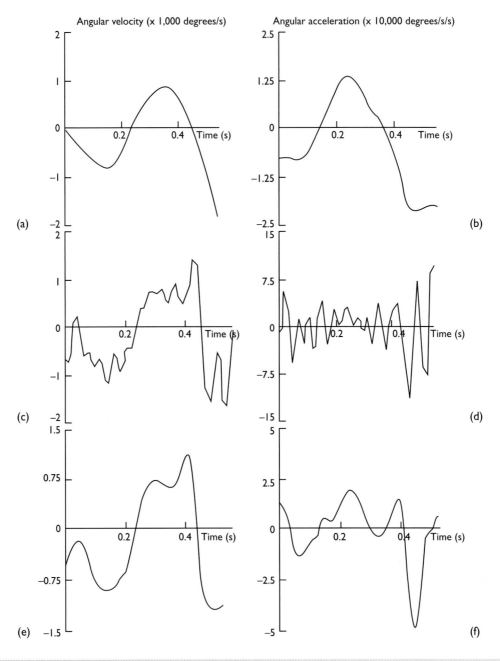

Quintic spline curve fitting

Many techniques used for the smoothing and differentiation of data in sports bio-mechanics involve the use of spline functions. These are a series of polynomial curves joined – or pieced – together at points called knots. This smoothing technique, which is performed in the time domain, can be considered to be the numerical equivalent of drawing a smooth curve through the data points. Indeed, the name 'spline' is derived from the flexible strip of rubber or wood used by draftsmen for drawing curves. Splines are claimed to represent the smoothness of human movement while rejecting the normally-distributed random noise in the digitised coordinates. Many spline techniques have a knot at each data point, obviating the need for the user to choose optimal knot positions. The user has simply to specify a weighting factor for each data point and select the value of the smoothing parameter, which controls the extent of the smoothing; generally the weighting factor should be the inverse of the estimate of the variance of the data point. This is easily established in sports biomechanics by repeated digitisation of a film or video sequence. The use of different weighting factors for different points can be useful, particularly if points are obscured from the camera and, therefore, have a greater variance than ones that can be seen clearly. Inappropriate choices of the smoothing parameter can cause problems of over-smoothing (Figures 4.19(a) and (b)) or under-smoothing (Figures 4.19(c) and (d)). The optimum smoothing is shown in Figures 4.19(e) and (f).

Generalised cross-validated quintic splines do not require the user to specify the error in the data to be smoothed, but instead automatically select an optimum smoothing parameter. Computer programs for spline smoothing are available in various software packages, such as MATLAB, and on the Internet (for example, http://isbweb.org.software/sigproc/gcvspl/gcvspl.fortran) and allow a choice of automatic or user-defined smoothing parameters. Generalised cross-validation can accommodate data points sampled at unequal time intervals. Splines can be differentiated analytically. Quintic splines are continuous up to the fourth derivative, which is a series of interconnected straight lines; this allows accurate generation of the second derivative, acceleration.

APPENDIX 4.2 BASIC VECTOR ALGEBRA

The vector in Figure 4.20(a) can be designated F (magnitude F) or OP (magnitude OP). Vectors can often be moved in space parallel to their original position, although some caution is necessary for force vectors (see Chapter 5). This allows easy graphical addition and subtraction, as in Figures 4.21(a) to (f). Note that the vector F in Figure 4.20(a) is equal in magnitude but opposite in direction to the vector G in Figure 4.20(b); if the direction of a vector is changed by 180°, the sign of the vector changes.

Figure 4.20 Vector representation: (a) vector **F** and (b) vector **G** = − **F**.

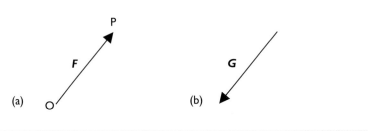

Vector addition and subtraction

When two or more vector quantities are added together the process is called 'vector composition'. Most vector quantities, including force, can be treated in this way. The single vector resulting from vector composition is known as the resultant vector or simply the resultant. In Figure 4.21 the vectors added are shown in black, the resultants in blue, and the graphical solutions for vector addition are shown between dashed vertical lines.

The addition of two or more vectors having the same direction results in a vector that has the same direction as the original vectors and a magnitude equal to the sum of the magnitudes of the vectors being added, as shown in Figure 4.21(a). If vectors directed in exactly opposite directions are added, the resultant has the direction of the longer vector and a magnitude that is equal to the difference in the magnitudes of the two original vectors, as in Figure 4.21(b).

When the vectors to be added lie in the same plane but not in the same or opposite directions, the resultant can be found using the vector triangle approach. The tail of the second vector is placed on the tip of the first vector. The resultant is then drawn from the tail of the first vector to the tip of the second, as in Figure 4.21(c). An alternative approach is to use a vector parallelogram, as in Figure 4.21(d). The vector triangle is more useful as it easily generalises to the vector polygon of Figure 4.21(e).

Although graphical addition of vectors is very easy for two-dimensional problems, the same is not true for three-dimensional ones. Component addition of vectors, using trigonometry, can be more easily generalised to the three-dimensional case. This technique is introduced for a two-dimensional example on pages 160–1.

The subtraction of one vector from another can be tackled graphically simply by treating the problem as one of addition, as in Figure 4.21(f); i.e., **J** = **F** − **G** is the same as **J** = **F** + (−**G**). Vector subtraction is often used to find the relative motion between two objects.

Figure 4.21 Vector addition: (a) with same direction; (b) in opposite directions; (c) using vector triangle; (d) using vector parallelogram; (e) using vector polygon; and (f) vector subtraction.

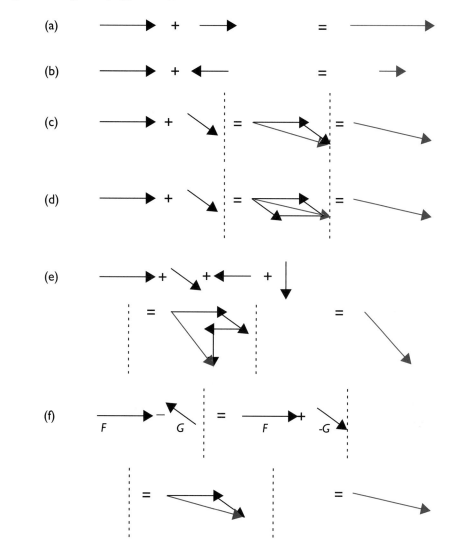

Vector resolution and components

Determining the perpendicular components of a vector is often useful in sports biomechanics. Examples include the normal and frictional components of ground reaction force, and the horizontal and vertical components of a projectile's velocity vector. It is, therefore, sometimes necessary to resolve a single vector into two or three perpendicular components – a process known as vector resolution. This is essentially the reverse of vector composition and can be achieved using a vector parallelogram or vector triangle. This is illustrated by the examples of Figure 4.22, where the blue arrows are the original vector and the black ones are its components. The vector parallelogram approach is more usual as the components then have a common origin, as in Figure 4.22(a).

Most angular motion vectors obey the rules of resolution and composition, but this is not true for angular displacements.

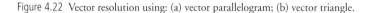

Figure 4.22 Vector resolution using: (a) vector parallelogram; (b) vector triangle.

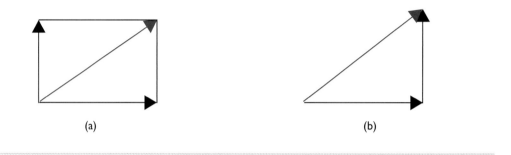

(a)   (b)

Vector addition and subtraction using vector components

Vector addition and subtraction can also be performed on the components of the vector using the rules of simple trigonometry. For example, consider the addition of the three vectors represented in Figure 4.23(a). Vector A is a horizontal vector with a magnitude (proportional to its length) of 1 unit. Vector B is a vertical vector of magnitude -2 units (it points vertically downwards, hence it is negative). Vector C has a magnitude of 3 units and is $120°$ measured anticlockwise from a right-facing horizontal line.

The components of the three vectors are summarised below (see also Figure 4.23(b)). The components (R_x, R_y) of the resultant vector $R = A + B + C$ are shown in Figure 4.23(c). The magnitude of the resultant is then obtained from the magnitudes of its two components, using Pythagoras' theorem $(R^2 = R_x^2 + R_y^2)$, as $R = (0.25 + 0.36)^{1/2} = 0.78$. Its direction to the right horizontal is given by the angle θ, whose tangent is R_y/R_x. That is $\tan \theta = -1.2$, giving $\theta = 130°$. The resultant R is rotated anticlockwise $130°$ from the right horizontal, as shown in Figure 4.25(d). (A second solution for $\tan \theta = -1.2$ is $\theta = -50°$, which would have been the answer if R_x had been $+0.5$ and R_y had been -0.6).

Vector	Horizontal component (x)	Vertical component (y)
A	1	0
B	0	-2
C	$3 \cos 120° = -3 \cos 60° = -1.5$	$3 \sin 120° = 3 \cos 60° = 2.6$
$A + B + C$	-0.5	0.6

Figure 4.23 Vector addition using components: (a) vectors to be added; (b) their components; (c) components of resultant; (d) resultant.

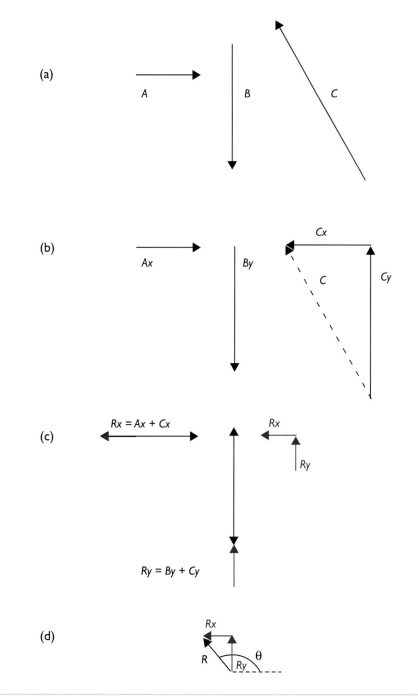

Vector multiplication

The rules for vector multiplication do not follow the simple algebraic rules for multiplying scalars.

Vector (cross) product

The vector (or cross) product of two vectors is useful in rotational motion because it enables, for example, angular motion vectors to be related to translational motion vectors (see below). It will be stated here in its simplest case for two vectors at right angles, as in Figure 4.24. The vector product of two vectors p and q inclined to one another at right angles is defined as a vector $p \times q$ of magnitude equal to the product $(p\,q)$ of the magnitudes of the two given vectors. Its direction is perpendicular to both vectors p and q in the direction in which the thumb points if the curled fingers of the right hand point from p to q through the right angle between them.

Figure 4.24 Vector cross-product.

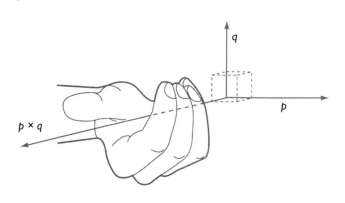

Scalar (dot) product

The scalar (dot) product of two vectors can be used, for example, to calculate power (a scalar, P) from force (F) and velocity (v), which are both vectors, using $P = F.v$. The dot product is a scalar so has no directional property. It is calculated as the product of the magnitudes of the two vectors and the cosine of the angle between them. It can also be calculated from the components of the vectors. For example, if $F = F_x + F_y$ and $v = v_x + v_y$ then $P = (F_x + F_y).(v_x + v_y)$. Now, the angle between the force and velocity along the same axis is $0°$, so the cosine of that angle is 1; however, the angle between the force and velocity along different axes is $90°$, so the cosine of that angle is 0. Hence, $P = F_x v_x + F_y v_y$. Please note that the power does NOT have x and y components, as it is a scalar (see also Figure 5.25(i)).

5 Causes of movement – forces and torques

Knowledge assumed
The importance of movement patterns for qualitative analysis (Chapters 1 to 3)
The body's cardinal planes and axes of movement (Chapter 1)
Simple algebraic manipulation (Chapter 4)
Quantitative analysis of sports movements (Chapter 4)
Basic vector algebra (Chapter 4)

INTRODUCTION

In Chapter 4, we covered quantitative videography of sports movements. In the final two chapters of this book, we will explore how these movements are generated. In this chapter, we will consider the forces that affect the movement of the sports performer. This branch of knowledge is often called 'kinetics', and is subdivided into linear and angular (rotational) kinetics; it deals with the action of forces and torques in producing or changing motion. We will also look at how we measure the forces and pressures acting on the sports performer.

BOX 5.1 LEARNING OUTCOMES

After reading this chapter you should be able to:

- appreciate how and why the variation with time of the forces acting on a sports performer can be viewed as another movement pattern that can be evaluated both qualitatively and quantitatively
- define force and identify the external forces acting in sport and how they affect movement
- understand the laws of linear kinetics and related concepts such as linear momentum
- calculate, from segmental and kinematic data, the position of the centre of mass of the human performer
- identify the ways in which rotation is acquired and controlled in sports movements
- understand the laws of angular kinetics and related concepts such as angular momentum
- appreciate why the measurement of the external forces acting on the sports performer is important in analysing sports movements
- understand the characteristics of a force plate that affect the accuracy of force measurement
- outline the procedures to be used when measuring force and pressure
- evaluate the information that can be obtained from force and pressure measurements.

FORCES IN SPORT

A force can be considered as the pushing or pulling action that one object exerts on another. Forces are vectors; they possess both a magnitude and a directional quality. The latter is specified by the direction in which the force acts and by the point on an object at which the force acts – its point of application. Alternatively, the total directional quality of the force can be given by its line of action, as in Figure 5.1.

The effects of a force are not altered by moving it along its line of action. Its effects on rotation – though not on linear motion – are changed if the force is moved parallel to the original direction and away from its line of action. A torque, also known as a

Figure 5.1 Directional quality of force.

moment of force or a turning effect, is then introduced; this is an effect tending to rotate the object (see below). A quantitative analyst should exercise care when solving systems of forces graphically and would usually adopt a vector approach (see Appendix 4.1).

The SI unit of force is the newton (N) and the symbol for a force vector is F. One newton is the force that when applied to a mass of one kilogram (1 kg), causes that mass to accelerate at 1 m/s^2 in the direction of the force application. A sports performer experiences forces both internal to and external to the body. Internal forces are generated by the muscles and transmitted by tendons, bones, ligaments and cartilage; these will be considered in Chapter 6. The main external forces, the combined effect of which determines the overall motion of the body, are as follows.

Weight

Weight is a familiar force (Figure 5.1) attributable to the gravitational pull of the Earth. It acts vertically downwards through the centre of gravity of an object towards the centre of the Earth. The centre of gravity (G in Figure 5.1) is an imaginary point at which the weight of an object can be considered to act. For the human performer, there is little difference between the positions of the centre of mass (see later) and the centre of gravity. The former is the term preferred in most modern sports biomechanics literature and will be used in the rest of this book. One reason for this preference is that the centre of gravity is a meaningless concept in weightless environments, such as space shuttles. An athlete with a mass of 50 kg has a weight (G) of about 490 N at sea level, at which the standard value of gravitational acceleration, g, is assumed to be 9.81 m/s^2.

Reaction forces

Reaction forces are the forces that the ground or other external surface exerts on the sports performer as a reaction to the force that the performer exerts on the ground or surface. This principle is known as Newton's third law of linear motion or the law of action–reaction. The vertical component of force acting on a person performing a standing vertical jump – with no arm action – is shown as a function of time in Figure 5.2. This movement pattern shows the period during which the jumper is on the ground (A), then in the air (B), and finally during landing (C). The component of the reaction force tangential to the surface, known as friction or traction, is crucially important in sport and is considered in the next section.

Friction

The ground, or other, contact force acting on an athlete (Figure 5.3(a)) can be resolved into two components, one (F_n) normal and one (F_t) tangential to the contact surface (Figure 5.3(b)). The former component is the normal force and the latter is the friction,

Figure 5.2 Vertical component of ground reaction force in a standing vertical jump with no arm action: (A) on the ground before take-off for the jump; (B) in the air; (C) landing.

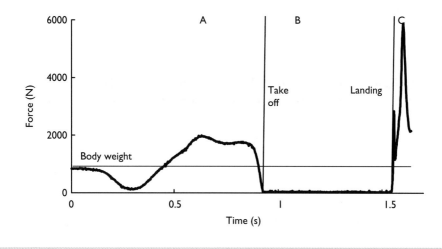

Figure 5.3 (a) Ground reaction force and (b) its components.

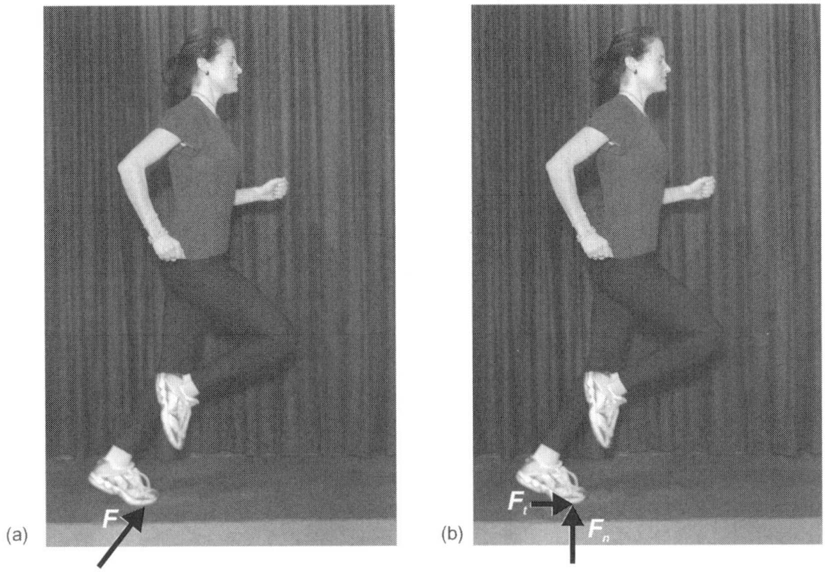

(a)

(b)

or traction, force. Traction is the term used when the force is generated by interlocking of the contacting objects, such as spikes penetrating a Tartan track; this interaction between objects is known as form locking. In friction, the force is generated by force locking, in which no surface penetration occurs. Without friction or traction, movement in sport would be very difficult.

If an object, such as a training shoe (Figure 5.4(a)), is placed on a sports surface material such as Tartan, it is possible to investigate how the friction force changes. The forces acting on the plane are shown in the 'free body diagram' of the shoe removed from its surroundings, but showing the forces that the surroundings exert on the shoe (Figure 5.4(b)). Because the shoe is not moving, these forces are in equilibrium. Resolving the weight of the shoe (G) along (F_t) and normal (F_n) to the plane, the magnitudes of the components are, respectively: $F_t = G \sin\theta$; $F_n = G \cos\theta$. Dividing F_t by F_n, we get $F_t/F_n = \tan\theta$. If the angle of inclination of the plane (θ) is increased, the friction force will eventually be unable to resist the component of the shoe's weight down the slope and the shoe will begin to slide. The ratio of F_t/F_n (= $\tan\theta$) at which this occurs is called the coefficient of (limiting) static friction (μ_s). The maximum sliding friction force that can be transmitted between two bodies is: $F_{t\ max} = \mu_s\ F_n$. This is known as Newton's law of friction and refers to static friction, just before there is any relative movement between the two surfaces. It also relates to conditions in which only the friction force prevents relative movement. For such conditions, the maximum friction force depends only on the magnitude of the normal force pressing the surfaces together and the coefficient of static friction (μ_s). This coefficient depends only on the

167

Figure 5.4 (a) Training shoe on an inclined plane and (b) its free body diagram.

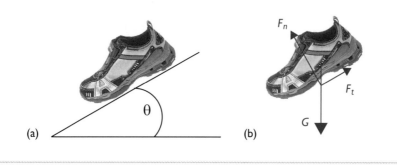

materials and nature (such as roughness) of the contacting surfaces and is, to a large extent, independent of the area of contact. The coefficient of friction should exceed 0.4 for safe walking on normal floors, 1.1 for running and 1.2 for all track and field events.

In certain sports in which spikes or studs penetrate or substantially deform a surface, the tangential (usually horizontal) force is transmitted by interlocking surfaces – traction – rather than by friction. Form locking then generates the tangential force, which is usually greater than that obtainable from static friction. For such force generation, a 'traction coefficient' can be defined similarly to the friction coefficient above.

Once the two surfaces are moving relative to one another, as when skis slide over snow, the friction force between them decreases and a 'coefficient of kinetic friction' is defined such that: $F_t = \mu_k F_n$, noting $\mu_k < \mu_s$. This coefficient is relatively constant up to a speed of about 10 m/s. Kinetic friction always opposes relative sliding motion between two surfaces.

Friction not only affects translational motion, it also influences rotation, such as when swinging around a high bar or pivoting on the spot. At present there is no agreed definition of, nor agreed method of measuring, rotational friction coefficients in sport. Frictional resistance also occurs when one object tends to rotate or roll along another, as for a hockey ball rolling across an AstroTurf™ pitch. In such cases it is possible to define a 'coefficient of rolling friction'. The resistance to rolling is considerably less than the resistance to sliding and can be established by allowing a ball to roll down a slope from a fixed height (1 m is often used) and then measuring the horizontal distance that it rolls on the surface of interest. In general, for sports balls rolling on sports surfaces, the coefficient of rolling friction is around 0.1.

Reducing friction

To reduce friction or traction between two surfaces it is necessary to reduce the normal force or the coefficient of friction. In sport, the latter can be done by changing the materials of contact, and the former by movement technique. Such a technique is known as unweighting, in which the performer imparts a downward acceleration to his or her centre of mass (Figure 5.5(a)), thus reducing the normal ground contact force to below body weight (Figure 5.5(b)). This technique is used, for example, in some turning skills in skiing and is often used to facilitate rotational movements.

Large coefficients of friction are detrimental when speed is wanted and friction opposes this. In skiing straight runs, kinetic friction is minimised by treating the base of the skis with wax. This can reduce the coefficient of friction to below 0.1. At the high speeds associated with skiing, frictional melting occurs, which further reduces the coefficient of friction to as low as 0.02 at speeds above 5 m/s. The friction coefficient is also affected by the condition of the snow–ice surface. In ice hockey and in figure and speed skating, the sharpened blades minimise the friction coefficient in the direction parallel to the blade length. The high pressures involved cause localised melting of the ice which, along with the smooth blade surface, reduces friction. Within the human body, where friction causes wear, synovial membranes of one form or another excrete synovial fluid to lubricate the structures involved, resulting in frictional coefficients as low as 0.001. This occurs in synovial joints, in synovial sheaths (such as that of the biceps brachii long head), and synovial sacs and bursae that protect tendons, for example at the tendon of quadriceps femoris near the patella.

Increasing friction

A large coefficient of friction or traction is often needed to permit quick changes of velocity or large accelerations. To increase friction, it is necessary to increase either the normal force or the friction coefficient. The normal force can be increased by weighting, the opposite process to unweighting. Other examples of increasing the

Figure 5.5 Unweighting: (a) forces acting on jumper; (b) force platform record.

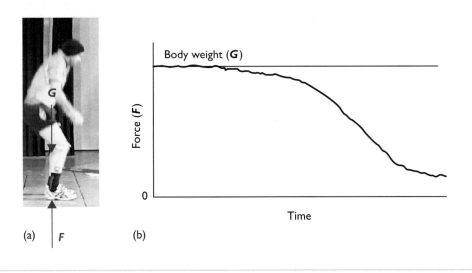

normal force include the use of inverted aerofoils on racing cars and the technique used by skilled rock and mountain climbers of leaning away from the rock face. To increase the coefficient of friction, a change of materials, conditions or locking mechanism is necessary. The last of these can lead to a larger traction coefficient through the use of spikes and, to a lesser extent, studs in for instance javelin throwing and running, in which large velocity changes occur.

Starting, stopping and turning

As noted above, form locking is very effective when large accelerations are needed. However, the use of spikes or studs makes rotation more difficult. This requires a compromise in general games, in which stopping and starting and rapid direction changes are combined with turning manoeuvres necessitating sliding of the shoe on the ground. When starting, the coefficient of friction or traction limits performance. A larger value on synthetic tracks compared with cinders, about 0.85 compared with 0.65, allows the runner a greater forward inclination of the trunk and a more horizontally directed leg drive and horizontal impulse. The use of starting blocks has similar benefits, with form locking replacing force locking. With lower values of the coefficient of friction, the runner must accommodate by using shorter strides. Although spikes can increase traction, energy is necessary to pull them out of the surface and it is questionable whether they confer any benefit when running on dry, dust-free synthetic tracks. When stopping, a sliding phase can be beneficial, unless a firm anchoring of the foot is required, as in the delivery stride of javelin throwing. Sliding is possible on cinder tracks and on grass and other natural surfaces; this potential is used by the grass and clay court tennis player. On synthetic surfaces, larger friction coefficients limit sliding; an athlete has to accommodate by unweighting by flexion of the knee. When turning, a large coefficient of friction requires substantial unweighting to permit a reduction in the torque needed to generate the turn. Without this unweighting, the energy demands of turning increase with rotational friction. In turning and stopping techniques in skiing, force locking is replaced by form locking using the edges of the skis; this substantially increases the force on the skis.

Effects of contact materials

The largest friction coefficients occur between two absolutely smooth, dry surfaces, owing to microscopic force locking at an atomic or molecular level. Many dry, clean, smooth metal surfaces in a vacuum adhere when they meet; in most cases, attempts to slide one past the other produce complete seizure. This perfect smoothness is made use of with certain rubbers in rock climbing shoes and racing tyres for dry surfaces. In the former case, the soft, smooth rubber adheres to the surface of the rock; in the latter, localised melting of the rubber occurs at the road surface. However in wet or very dusty conditions the loss of friction is substantial, as adhesion between the rubber and surface is prevented. The coefficient of friction reduces for a smooth tyre from around 5.0 on a dry surface to around 0.1 on a wet one. A compromise is afforded by treaded tyres with dry and wet surface friction coefficients of around 1.0 and 0.4, respectively. The treads allow water to be removed from the contact area between the tyre and road surface. Likewise, most sports shoes have treaded or cleated soles, and club and racket grips are rarely perfectly smooth. In some cases, the cleats on the sole of a sports shoe will also provide some form locking with certain surfaces.

Pulley friction

Passing a rope around the surface of a pulley makes it easier to resist a force of large magnitude at one end by a much smaller force at the other end because of the friction between the rope and the pulley. This principle is used, for example, in abseiling techniques in rock climbing and mountaineering. It also explains the need for synovial membranes to prevent large friction forces when tendons pass over bony prominences.

Buoyancy

Buoyancy is the force experienced by an object immersed, or partly immersed, in a fluid. It always acts vertically upwards at the centre of buoyancy (CB in Figure 5.6). The magnitude of the buoyancy force (B) is expressed by Archimedes' principle, 'the upthrust is equal to the weight of fluid displaced', and is given by $B = V \rho g$, where V is the volume of fluid displaced, ρ is the density of the fluid and g is gravitational acceleration. The buoyancy force is large in water – pure fresh water has a density of 1000 kg/m^3 – and much smaller, but not entirely negligible, in air, which has a density of around 1.23 kg/m^3. For a person or an object to float and not sink, the magnitudes of the buoyancy force and the weight of the object must be equal, so that $B = G$.

The swimmer in Figure 5.6 will only float if her average body density is less than or equal to the density of water. How much of her body is submerged will depend on the ratio of the two densities. If the density of the swimmer is greater than that of the water,

Figure 5.6 Buoyancy force: (a) forces acting; (b) forces in equilibrium.

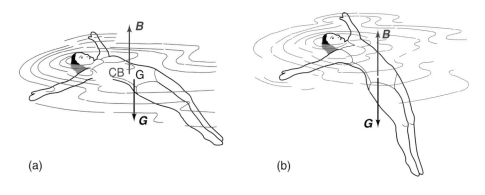

(a)   (b)

she will sink. It is easier to float in sea water, which has a density of around 1020 kg/m^3, than in fresh water. For the human body, the relative proportions of tissues will determine whether sinking or floating occurs. Typical densities for body tissues are: fat, 960 kg/m^3; muscle, 1040–1090 kg/m^3; bone, 1100 (cancellous) to 1800 kg/m^3 (compact). The amount of air in the lungs is also very important. Most Caucasians can float with full inhalation whereas most Negroes cannot float even with full inhalation because of their different body composition, which is surely a factor contributing to the shortage of world-class black swimmers. Most people cannot float with full exhalation. Women float better than men because of an inherently higher proportion of body fat and champion swimmers have, not surprisingly, higher proportions of body fat than other elite athletes.

Fluid dynamic forces

Basic fluid mechanics

All sports take place within a fluid environment; the fluid is air in running, liquid in underwater turns in swimming, or both, for example in sailing. Unlike solids, fluids flow freely and their shape is only retained if enclosed in a container; particles of the fluid alter their relative positions whenever a force acts. Liquids have a volume that stays the same while the shape changes. Gases expand to fill the whole volume available by changing density. This ability of fluids to distort continuously is vital to sports motions as it permits movement. Although fluids flow freely, there is a resistance to this flow known as the 'viscosity' of the fluid. Viscosity is a property causing shear stresses between adjacent layers of moving fluid, leading to a resistance to motion through the fluid.

In general, the instantaneous velocity of a small element of fluid will depend both on time and its spatial position. A small element of fluid will generally follow a complex path known as the path line of the element of fluid. An imaginary line that lies tangential to the direction of flow of the fluid particles at any instant is called a streamline; streamlines have no fluid flow across them.

An important principle in fluid dynamics in sport is Bernoulli's principle, which, in essence, states that reducing the flow area, as for fluid flow past a runner, results in an increase in fluid speed and a decrease in fluid pressure. There are two very important ratios of fluid forces in sport. The 'Reynolds number' is important in all fluid flow in sport; it is the ratio of the inertial force in the fluid to the viscous force, and is calculated as $v\,l\,/\,\upsilon$ where v is a 'characteristic speed' (usually the relative velocity between the fluid and an object), l is some 'characteristic length' of the object and υ is the 'kinematic viscosity' of the fluid. The 'Froude number' is the square root of the ratio of inertial to gravity forces; it is important whenever an interface between two fluids occurs and waves are generated, as happens in almost all water sports. The Froude number is calculated as $v/\sqrt{(l\,g)}$ where v and l are the characteristic speed and length, as above, and g is gravitational acceleration.

Laminar and turbulent flow

In laminar flow, the fluid particles move only in the direction of the flow. The fluid can be considered to consist of discrete plates (or laminae) flowing past one another. Laminar flow occurs at moderate speeds past objects of small diameter, such as a table tennis ball, and energy is exchanged only between adjacent layers of the flowing fluid. Turbulent flow is predominant in sport. The particles of fluid have fluctuating velocity components in both the main flow direction and perpendicular to it. Turbulent flow is best thought of as a random collection of rotating eddies or vortices; energy is exchanged by these turbulent eddies on a greater scale than in laminar flow.

The type of flow that exists under given conditions depends on the Reynolds number. For low Reynolds numbers the fluid flow is laminar. At the 'critical' Reynolds number, the flow passes through a transition region and then becomes turbulent at a slightly higher Reynolds number. The Reynolds number at which flow changes from laminar to turbulent depends very much on the object past which the fluid flows. Consider a fairly flat boat hull moving through stationary water. Let us use the distance along the hull from the bow as the 'characteristic length' for the 'local' Reynolds number of the water flow past that point on the hull. Furthermore, we will define the characteristic speed as that of the boat moving through the water. For this example, the critical Reynolds number is in the range 100 000–3 000 000, depending on the nature of the flow in the water away from the boat and the surface roughness of the hull. The fluid flow close to the hull (in the 'boundary layer' discussed below) will change from laminar to turbulent at the point along the hull where the 'local' Reynolds number equals the critical value. If this does not happen, the flow will remain laminar along the whole length of the hull. For flow past a ball, the ball diameter is used as the characteristic length and the characteristic speed is the speed of the ball relative to the air. The critical Reynolds number, depending on ball roughness and flow conditions outside the boundary layer, is in the range 100 000–300 000.

The boundary layer

When relative motion occurs between a fluid and an object, as for flow of air or water past the sport performer, the fluid nearest the object is slowed down because of its viscosity. The region of fluid affected in this way is known as the boundary layer. Within this layer, the relative velocity of the fluid and object changes from zero at the surface of the object to the free stream velocity, which is the difference between the velocity of the object and the velocity of the fluid outside the boundary layer. The slowing down of the fluid is accentuated if the flow of the fluid is from a wider to a narrower cross-section of the object, as the fluid is then trying to flow from a low-pressure region to a high-pressure region. Some of the fluid in the boundary layer may lose all its kinetic energy. The boundary layer then separates from the body at the separation points (S in Figure 5.7), leaving a low-pressure area, known as the wake, behind the object.

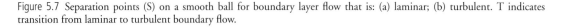

Figure 5.7 Separation points (S) on a smooth ball for boundary layer flow that is: (a) laminar; (b) turbulent. T indicates transition from laminar to turbulent boundary flow.

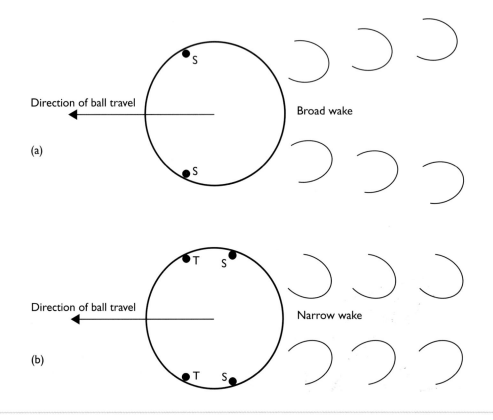

An object moving from a region of low pressure to one of high pressure, experiences a drag force, which will be discussed below. The boundary layer separation, which leads to the formation of the wake, occurs far more readily if the fluid flow in the boundary layer is laminar (Figure 5.7(a)) than if the flow is turbulent (Figure 5.7(b)). This is because kinetic energy is more evenly distributed across a turbulent boundary layer, enabling the fluid particles near the boundary to better resist the increasing pressure. Figure 5.7 shows the difference in the separation point (S) positions and the size of the wake between laminar and turbulent boundary layers on a ball. The change from laminar to turbulent boundary layer flow will occur, for a given object and conditions, at a speed related to the critical Reynolds number. The change occurs at the transition point (T, Figure 5.7(b)), and the relationship between the transition and separation points is very important. If separation occurs before its transition to turbulent flow, a large wake is formed (Figure 5.7(a)), whereas transition to turbulent flow upstream of the separation point results in a smaller wake (Figure 5.7(b)).

Drag forces

If an object is symmetrical with respect to the fluid flow, such as a non-spinning soccer ball, the fluid dynamic force acts in the direction opposite to the motion of the object and is termed a drag force. Drag forces resist motion and, therefore, generally restrict sports performance. They can, however, have beneficial propulsive effects, as in swimming and rowing. To maintain a runner in motion at a constant speed against a drag force requires an expenditure of energy equal to the product of the drag force and the speed. If no such energy is present, as for a projectile, the object will decelerate at a rate proportional to the area presented by the object to the fluid flow (the so-called 'frontal' area, A) and inversely proportional to the mass of the body, m. The mass to area ratio (m/A) is crucial in determining the effect that air resistance has on projectile motion. A shot, with a very high ratio of mass to area, is hardly affected by air resistance whereas a cricket ball, with only 1/16th the mass to area ratio of the shot, is far more affected. A table tennis ball (1/250th the mass to area ratio of the shot) has a greatly altered trajectory.

Pressure drag

Pressure drag, or wake drag, contributes to the fluid resistance experienced by, for example, projectiles and runners. This is the major drag force in most sports and is caused by boundary layer separation leaving a low-pressure wake behind the object. The object, tending to move from a low-pressure to a high-pressure region, experiences a drag force. The pressure drag can be reduced by minimising the disturbance that the object causes to the fluid flow, a process known as 'streamlining'. An oval shape, similar to a rugby ball, has only two-thirds of the pressure drag of a spherical ball with the same frontal area. The pressure drag is very small on a streamlined aerofoil shape, such as the cross-section of a glider wing. Streamlining is very important in motor car and motor cycle racing, and in discus and javelin throwing. Swimmers and skiers can reduce the pressure drag forces acting on them by adopting streamlined shapes. The adoption of a streamlined shape is of considerable advantage to downhill skiers.

If we increase the speed of an object through a fluid – such as a ball through the air – we find a dramatic change in the drag as the boundary layer flow changes from laminar to turbulent. As this transition occurs at the critical Reynolds number, the drag decreases by about 65%. Promoting a turbulent boundary layer is an important mechanism in reducing pressure drag if the speed is close to the value necessary to achieve the critical Reynolds number. At such speeds, which are common in ball sports, roughening the surface promotes turbulence in the boundary layer, encouraging this decrease in drag. The nap of tennis balls and the dimples on a golf ball are examples of roughness helping to induce boundary layer transition, thereby reducing drag. Within the Reynolds number range 110 000–175 000, which corresponds to ball speeds off the tee of 45–70 m/s, the dimples on a golf ball cause the drag coefficient to decrease proportionally to speed. The drag force then increases only proportionally to speed, rather than speed squared, benefiting the hard-hitting player.

Many sport balls are not uniformly rough. Then, within a speed range somewhat below the critical Reynolds number, it is possible for roughness elements on one part of the ball to stimulate transition of the boundary layer to turbulent flow, while the boundary layer flow on the remaining smoother portion of the ball remains laminar. This is very important, for example, in cricket ball swing, in which the asymmetrical disposition of the ball's seam accounts for the lateral movement of the ball known as swing. The seam promotes turbulence in the boundary layer on the 'rough' side of the ball, on which the seam is upstream of the separation point, as in Figure 5.7(b), while separation occurs on the other ('smooth') side of the ball, as in Figure 5.7(a). The asymmetrical wake causes the ball to swing towards the side to which the seam points. For reverse swing to occur, the ball must be released above the critical speed for the smooth side of the ball, which can only be done by bowlers who can achieve such speeds. The boundary layer becomes turbulent on both hemispheres before separation. On the rough side, the turbulent boundary layer thickens more rapidly and separates earlier than on the smooth side. The result is the reversal of the directions of wake displacement and, therefore, swing.

Skin friction drag
Skin friction drag is the force caused by friction between the molecules of fluid and a solid boundary. It is only important for streamlined bodies for which separation – and pressure drag – has been minimised. Unlike pressure drag, skin friction drag is reduced by having a laminar as opposed to a turbulent boundary layer. This occurs because the rate of shear at the solid boundary is greater for turbulent flow. Reduction of skin friction drag is important for racing cars, racing motor cycles, gliders, hulls of boats, skiers and ski-jumpers and, perhaps, swimmers. It is minimised by reducing the roughness of the surfaces in contact with the fluid.

Wave drag
Wave drag occurs only in sports in which an object moves through both water and air. As the object moves through the water, the pressure differences at its boundary cause the water level to rise and fall and waves are generated. The energy of the waves is provided by the object, which experiences a resistance to its motion. The greater the speed of the body, the larger the wave drag, which is important in most aquatic sports. Wave drag also depends on the wave patterns generated and the dimensions of the object. The drag is often expressed as a function of the Froude number. Speed boats and racing yachts are designed to plane – to ride high in the water – at their highest speeds so that wave drag – and pressure drag – are then very small. In swimming the wave drag is small compared with the pressure drag, unless the swimmer's speed is above about 1.6 m/s, when a bow wave is formed.

Other forms of drag
Spray-making drag occurs in some water sports because of the energy involved in generating spray. It is usually negligible, except perhaps during high-speed turns in surfing and windsurfing. Induced drag arises from a three-dimensional object that is generating lift. It can be minimised by having a large aspect ratio – the ratio of the

dimension of the object perpendicular to the flow direction to the dimension along the flow direction. Long, thin wings on gliders minimise the induced drag whereas a javelin has entirely the wrong shape for this purpose.

Lift forces

If an asymmetry exists in the fluid flow around a body, the fluid dynamic force will act at some angle to the direction of motion and can be resolved into two component forces. These are a drag force opposite to the flow direction and a lift force perpendicular to the flow direction. Such asymmetry may be caused in three ways. For a discus and javelin, for example, it arises from an inclination of an axis of symmetry of the body to the direction of flow (Figure 5.8(a)). Another cause is asymmetry of the body (Figure 5.8(b)); this is the case for sails, which act similarly to the aerofoil-shaped wings of an aircraft, and the hands of a swimmer, which function as hydrofoils. The Magnus effect (Figure 5.8(c)) occurs when rotation of a symmetrical body, such as a ball, produces asymmetry in the fluid flow.

Swimmers use a mixture of lift and drag forces for propulsion, with their hands acting as rudimentary hydrofoils. A side view of a typical path of a front crawl swimmer's hand relative to the water is shown in Figure 5.9(a). In the initial and final portions of the pull phase of this stroke (marked 'i' and 'f' in Figure 5.9(a)) only the lift force (L) can make a significant contribution to propulsion. In the middle region of the pull phase (marked 'm'), the side view would suggest that drag (D) is the dominant contributor. However, a view from in front of the swimmer (Figure 5.9(b)) or below (Figure 5.9(c)) shows an S-shaped pull pattern in the sideways plane. These sideways movements of the hands generate significant propulsion through lift forces perpendicular to the path of travel of the hand (Figure 5.9(c)). Swimmers need to develop a 'feel' for the water flow over their hands, and to vary the hand 'pitch' angle with respect to the flow of water to optimise the propulsive forces throughout the stroke. The shape of both oars and paddles suggests that they also can behave as hydrofoils. In rowing, for example, the propulsive forces generated are a combination of both lift and drag components and not just drag. The velocity of the oar or paddle relative to the water is the crucial factor. If this is forwards at any time when the oar is submerged, the drag is in the wrong direction to provide propulsion. Only a lift force can fulfil this function (see Figure 5.9). The wing paddle in kayaking was developed to exploit this propulsive lift effect.

Many ball sports involve a spinning ball. Consider, for example, a ball moving through a fluid, and having backspin as in Figure 5.8(c). The top of the ball is moving in the same direction as the air relative to the ball, while the bottom of the ball is moving against the air stream. The rotational motion of the ball is transferred to the thin boundary layer adjacent to the surface of the ball. On the upper surface of the ball this 'circulation' imparted to the boundary layer reduces the difference in velocity across the boundary layer and delays separation. On the lower surface of the ball, the boundary layer is moving against the rest of the fluid flow, known as the free stream.

Figure 5.8 Generation of lift: (a) inclination of an axis of symmetry; (b) body asymmetry; (c) Magnus effect.

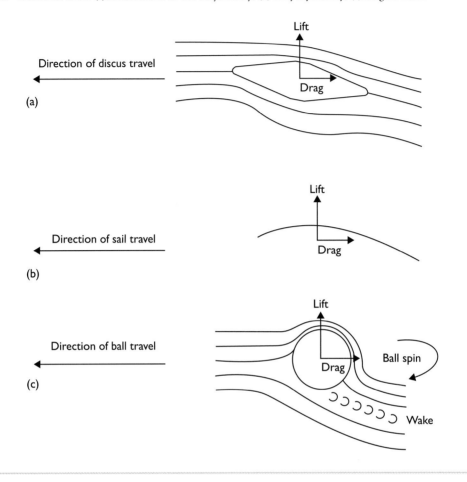

This increases the velocity difference across the boundary layer and separation still occurs. The resulting wake has, therefore, been deflected downwards, as can be seen in Figure 5.8(c). Newton's laws of motion imply that the wake deflection is due to a force provided by the ball acting downwards on the air and that a reaction moves the ball away from the wake. This phenomenon is known as the Magnus effect. For a ball with backspin, the force acts perpendicular to the motion of the ball – it is a lift force.

Golf clubs are lofted so that the ball is undercut, producing backspin. This varies from approximately 50 Hz for a wood to 160 Hz for a nine iron. The lift force generated can even be sufficient to cause the initial ball trajectory to be curved slightly upwards. The lift force increases with the spin and substantially increases the length of drive compared to that with no spin. The main function of golf ball dimples is to assist the transfer of the rotational motion of the ball to the boundary layer of air to increase the Magnus force and give optimum lift. Backspin and topspin are used in games such

Figure 5.9 Typical path of a swimmer's hand relative to the water: (a) side view; (b) view from in front of the swimmer; (c) view from below the swimmer.

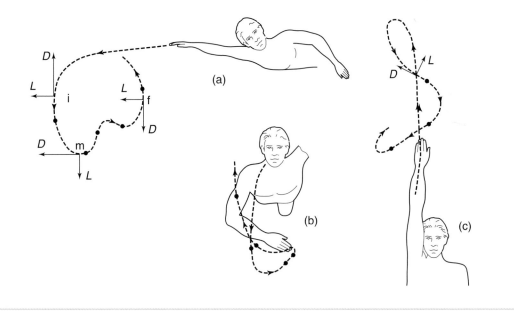

as tennis and table tennis both to vary the flight of the ball and to alter its bounce. In cricket, a spin bowler usually spins the ball so that it is rotating about the axis along which it is moving (its velocity vector), and the ball only deviates when it contacts the ground. However, if the ball's spin axis does not coincide with its velocity vector, the ball will also move laterally through the air. A baseball pitcher also uses the Magnus effect to 'curve' the ball; at a pitching speed of 30 m/s the spin imparted can be as high as 30 Hz, which gives a lateral deflection of 0.45 m in 18 m. A soccer ball can be made to swerve, or 'bend', in flight by moving the foot across the ball as it is kicked. This causes rotation of the ball about the vertical axis. If the foot is moved from right to left as the ball is kicked, the ball will swerve to the right. Slicing and hooking of a golf ball are caused, inadvertently, by sidespin imparted at impact. In crosswinds, the relative direction of motion between the air and the ball is changed; a small amount of sidespin, imparted by 'drawing' the ball with a slightly open club face, or 'fading' the ball with a slightly closed club face depending on the wind direction, can then increase the length of the drive.

A negative Magnus effect can also occur for a ball travelling below the critical Reynolds number. This happens when the boundary layer flow remains laminar on the side of the ball moving in the direction of the relative air flow, as the Reynolds number here remains below the critical value. On the other side of the ball, the rotation increases the relative speed between the air and the ball so that the boundary layer becomes turbulent. If this happens on a back-spinning ball, laminar boundary layer flow will occur on the top surface and turbulent on the bottom surface. The wake will be deflected upwards, the opposite from the normal Magnus effect discussed above, and the ball will plummet to the ground under the action of the negative lift force. Reynolds numbers in many ball sports are close to the critical value, and the negative Magnus effect may, therefore, be important. It has occasionally been proposed as an explanation for certain unusual ball behaviour in, for example, the 'floating' serve in volleyball.

Impact forces

Impact forces occur whenever two or more objects collide. They are usually very large and of short duration compared to other forces acting. The most important impact force for the sport and exercise participant is that between that person and some external object, for example a runner's foot striking a hard surface. These forces can be positive biologically as they can promote bone growth, providing that they are not too large; large impact forces are one factor that can increase the injury risk to an athlete. An example of an impact force is shown by the force peak just after the start of the landing phase (C) in Figure 5.2.

Impacts involving sports objects, such as a ball and the ground, can affect the technique of a sports performer. For example, the spin imparted by the server to a tennis ball will affect how it rebounds, which will influence the stroke played by the receiver. Impacts of this type are termed oblique impacts and involve, for example, a ball hitting the ground at an angle of other than 90°, as in a tennis serve, and a bat or racket hitting a moving ball. If the objects at impact are moving along the line joining their centres of mass, the impact is 'direct', as when a ball is dropped vertically on to the ground. Direct impacts are unusual in sport, in which oblique impacts predominate.

COMBINATIONS OF FORCES ON THE SPORTS PERFORMER

In sport, more than one external force usually acts on the performer and the effect produced by the combination of these forces will depend on their magnitudes and relative directions. Figure 5.10(a) shows a biomechanical system, here a runner, isolated from the surrounding world. The effects of those surroundings, which for the runner are weight and ground reaction force, are represented on the diagram as force vectors. As mentioned on page 167, such a diagram is known as a 'free body diagram', which should be used whenever carrying out a biomechanical analysis of 'force systems'. The effects of different types of force system can be considered as follows.

Statics is a very useful and mathematically simple and powerful branch of mechanics. It is used to study force systems in which the forces are in equilibrium, such that they have no resultant effect on the object on which they act, as in Figure 5.6(b). In this figure, the buoyancy force, B, and the weight of the swimmer, G, share the same line of action and are equal in magnitude but have opposite directions, so that $B = G$. This approach may seem to be somewhat limited in sport, in which the net, or resultant, effect of the forces acting is usually to cause the performer or object to accelerate, as in Figure 5.10(a). In this figure, the resultant force can be obtained by moving the ground reaction force, F, along its line of action, which passes through the centre of mass in this case, giving Figure 5.10(b). As the resultant force passes through the runner's centre of mass, the runner can be represented as a point, the centre of mass, at which the entire runner's mass is considered to be concentrated: only changes in linear motion will occur for such a force system. The resultant of F and G will be the net force acting on the runner. This net force equals the mass of the runner, m, multiplied by the acceleration, a, of the centre of mass. Symbolically, this is written as $F + G = m\mathbf{a}$. This is one form of Newton's second law of motion (see Box 5.2).

More generally, as in Figure 5.10(c), the resultant force will not act through the centre of mass; a torque – also called a moment of force – will then tend to cause the runner to rotate about his or her centre of mass. The magnitude of this torque about a point – here the centre of mass – is the product of the force and its moment arm, which is the perpendicular distance of the line of action of the force from that point. Rotation will be considered in detail later in this chapter.

It is possible to treat dynamic systems of forces, such as those represented in Figure 5.10, using the equations of static equilibrium. To do this, however, we need to introduce an imaginary force into the dynamic system, which is equal in magnitude to the resultant force but opposite in direction, to produce a quasi-static force system.

Figure 5.10 Forces on a runner: (a) free body diagram of dynamic force system; (b) resultant force; (c) free body diagram with force not through centre of mass.

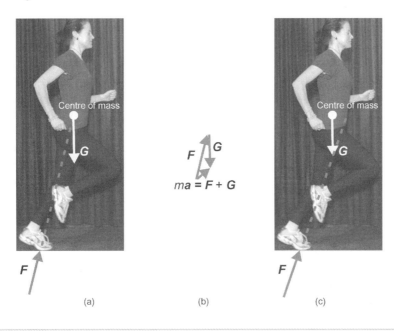

This imaginary force is known as an 'inertia' force. Its introduction allows the use of the general and very simple equations of static equilibrium for forces (F) and torques (M): $\Sigma F = 0$; $\Sigma M = 0$; that is, the vector sum (Σ) of all the forces, including the imaginary inertia forces, is zero and the vector sum of all the torques, including the imaginary inertia torques, is also zero (see Appendix 4.2 for a simple graphical method to calculate vector sums). These vector equations of static equilibrium can be applied to all force systems that are static or have been made quasi-static through the use of inertia forces. How the vector equations simplify to the scalar equations used to calculate the magnitudes of forces and moments of force will depend on how the forces combine to form the system of forces. Two-dimensional systems of forces are known as planar force systems and have forces acting in one plane only; three-dimensional force systems are known as spatial force systems. Force systems can be classified as follows:

- Linear (also called collinear) force systems consist of forces with the same line of action, such as the forces in a tug-of-war rope or the swimmer in Figure 5.6(b). No torque equilibrium equation is relevant for such systems as all the forces act along the same line.
- Concurrent force systems have the lines of action of the forces passing through a common point, such as the centre of mass. The collinear system in the previous paragraph is a special case. The runner in Figure 5.10(a) is an example of a planar concurrent force system and many spatial ones can also be found in sport and exercise movements. As all forces pass through the centre of mass, no torque equilibrium equation is relevant.
- Parallel force systems have the lines of action of the forces all parallel; they can be planar, as in Figure 5.6(a), or spatial. The tendency of the forces to rotate the object about some point means that the equation of moment equilibrium must be considered. The simple cases of first-class and third-class levers in the human musculoskeletal system are examples of planar parallel force systems and are shown in Figures 5.11(a) and (b). The moment equilibrium equation in these examples reduces to the 'principle of levers'. This states that the product of the magnitudes of the muscle force and its moment arm, sometimes called the force arm, equals the product of the resistance and its moment arm, called the resistance arm. Symbolically, using the notation of Figure 5.11: $F_m\, r_m = F_r\, r_r$. The force equilibrium equation leads to $F_j = F_m + F_r$ and $F_j = F_m - F_r$ for the joint force (F_j) in the first-class and third-class levers of Figures 5.11(a) and (b), respectively. It is worth mentioning here that the example of a second-class lever often quoted in sports biomechanics textbooks – that of a person rising onto the toes treating the floor as the fulcrum – is contrived. Few, if any, such levers exist in the human musculoskeletal system. This is not surprising as they represent a class of mechanical lever intended to enable a large force to be moved by a small one, as in a wheelbarrow. The human musculoskeletal system, by contrast, achieves speed and range of movement but requires relatively large muscle forces to accomplish this against resistance.
- Finally, general force systems may be planar or spatial, have none of the above simplifications and are the ones normally found in sports biomechanics, such as

Figure 5.11 Levers as examples of parallel force systems: (a) first-class lever; (b) third-class lever.

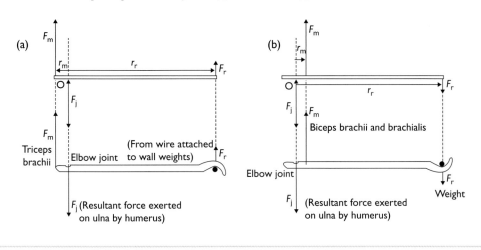

when analysing the various soft tissue forces acting on a body segment. These force systems will not be covered further in this book.

The vector equations of statics and the use of inertia forces can aid the analysis of the complex force systems that are commonplace in sport. However, many sports biomechanists feel that the use of the equations of static equilibrium obscures the dynamic nature of force in sport, and that it is more revealing to deal with the dynamic equations of motion, an approach that I prefer. In sport, force systems almost always change with time, as in Figure 5.2, which shows the vertical component of ground reaction force recorded from a force platform during a standing vertical jump. The effect of the force at any instant is reflected in an instantaneous acceleration of the performer's centre of mass. The change of the force with time determines how the velocity and displacement of the centre of mass change, and it is important to remember this.

MOMENTUM AND THE LAWS OF LINEAR MOTION

Inertia and mass

The inertia of an object is its reluctance to change its state of motion. Inertia is directly measured or expressed by the mass of the object, which is the quantity of matter of which the object is composed. It is more difficult to accelerate an object of large mass, such as a shot, than one of small mass, such as a dart. Mass is a scalar, having no directional quality; the SI unit of mass is the kilogram (kg).

Momentum

Linear momentum (more usually just called momentum) is the quantity of motion possessed by a particle or rigid body measured by the product of its mass and the velocity of its centre of mass. It is a very important quantity in sports biomechanics. As it is the product of a scalar and a vector, it is itself a vector whose direction is identical to that of the velocity vector. The unit of linear momentum is kg m/s.

BOX 5.2 NEWTON'S LAWS OF LINEAR MOTION

These laws completely determine the motion of a point, such as the centre of mass of a sports performer, and are named after the great British scientist of the sixteenth century, Sir Isaac Newton. They have to be modified to deal with the rotational motion of the body as a whole or a single body segment, as below. They have limited use when analysing complex motions of systems of rigid bodies, but these are beyond the scope of this book.

First law (law of inertia)

An object will continue in a state of rest or of uniform motion in a straight line (constant velocity) unless acted upon by external forces that are not in equilibrium; straight line skating is a close approximation to this state; a skater can glide across the ice at almost constant velocity as the coefficient of friction is so small. To change velocity, the blades of the skates need to be turned away from the direction of motion to increase the force acting on them. In the flight phase of a long jump the horizontal velocity of the jumper remains almost constant, as air resistance is small. However, the vertical velocity of the jumper changes continuously because of the jumper's weight – an external force caused by the gravitational pull of the Earth.

Second law (law of momentum)

The rate of change of momentum of an object is proportional to the force causing it and takes place in the direction in which the force acts. For an object of constant mass such as the human performer, this law simplifies to: the mass multiplied by the acceleration of that mass is equal to the force acting. When a ball is kicked, in soccer for example, the acceleration of the ball will be proportional to the force applied to the ball by the kicker's foot and inversely proportional to the mass of the ball.

Third law (law of interaction)

For every action, or force, exerted by one object on a second, there is an equal and opposite force, or reaction, exerted by the second object on the first. The ground reaction force experienced by the runner of Figure 5.3 is equal and opposite to the force exerted by the runner on the ground (F); this latter force would be shown on a free body diagram of the ground.

Impulse of a force

Newton's second law of linear motion (the law of momentum) can be expressed symbolically at any time, t, as $\boldsymbol{F} = \mathrm{d}\boldsymbol{p}/\mathrm{d}t = \mathrm{d}(m\,\boldsymbol{v})/\mathrm{d}t$. That is, $\boldsymbol{F}$, the net external force acting on the body, equals the rate of change ($\mathrm{d}/\mathrm{d}t$) of momentum ($\boldsymbol{p} = m\,\boldsymbol{v}$). For an object of constant mass (m), this becomes: $\boldsymbol{F} = \mathrm{d}\boldsymbol{p}/\mathrm{d}t = m\,\mathrm{d}\boldsymbol{v}/\mathrm{d}t = m\,\boldsymbol{a}$, where v is velocity and a is acceleration. If we now sum these symbolic equations over a time interval we can write: $\int \boldsymbol{F}\,\mathrm{d}t = \int \mathrm{d}(m\,\boldsymbol{v})$; this equals $m\int \mathrm{d}\boldsymbol{v}$, if m is constant. The symbol $\int$ is called an integral, which is basically the summing of instantaneous forces. The left side, $\int \boldsymbol{F}\,\mathrm{d}t$, of this equation is the impulse of the force, for which the SI unit is newton-seconds, N s. This impulse equals the change of momentum of the object ($\int \mathrm{d}(m\,\boldsymbol{v})$ or $m\int \mathrm{d}\boldsymbol{v}$ if m is constant). This equation is known as the impulse–momentum equation and, with its equivalent form for rotation, is an important foundation of studies of human dynamics in sport. The impulse is the area under the force–time curve over the time interval of interest and can be calculated graphically or numerically. The impulse of force can be found from a force–time pattern, such as Figure 5.2 or 5.12, which is easily obtained from a force platform.

The impulse–momentum equation can be rewritten for an object of constant mass (m) as $F\Delta t = m\,\Delta v$, where F is the mean value of the force acting during a time interval Δt during which the speed of the object changes by Δv (the Greek symbol delta, Δ, simply designates a change). The change in the horizontal velocity of a sprinter from the gun firing to leaving the blocks depends on the horizontal impulse of the force exerted by the sprinter on the blocks (from the second law of linear motion) and is inversely proportional to the mass of the sprinter. In turn, the impulse of the force exerted by the blocks on the sprinter is equal in magnitude but opposite in direction to that exerted, by muscular action, by the sprinter on the blocks (from the third law of linear motion). Obviously, a large horizontal velocity off the blocks is desirable. However, a compromise is needed as the time spent in achieving the required impulse (Δt) adds to the time spent running after leaving the blocks to give the recorded race time. The production of a large impulse of force is also important in many sports techniques of hitting, kicking and throwing to maximise the speed of the object involved. In javelin throwing, for example, the release speed of the javelin depends on the impulse applied to the javelin by the thrower during the delivery stride and the impulse applied by ground reaction and gravity forces to the combined thrower–javelin system throughout the preceding phases of the throw. In catching a ball, the impulse required to stop the ball is determined by the mass (m) and change in speed (Δv) of the ball. The catcher can reduce the mean force (F) acting on his or her hands by increasing the duration of the contact time (Δt) by 'giving' with the ball.

FORCE–TIME GRAPHS AS MOVEMENT PATTERNS

Consider an international volleyball coach who wishes to assess the vertical jumping capabilities of his or her squad members. Assume that this coach has access to a force plate, a device that records the variation with time of the contact force between a person and the surroundings (see below). The coach uses the force plate to record the force–time graph exerted by the players performing standing vertical jumps (see, for example, Figure 5.2). These graphs provide another movement pattern for both the qualitative and the quantitative movement analyst; our world is exceedingly rich in such patterns. Each player's force–time graph, after subtracting his or her weight, is easily converted to an acceleration–time graph (Figure 5.12(a)) as acceleration equals force divided by the player's mass, where the acceleration is that of the jumper's centre of mass.

Qualitative evaluation of a force–time or acceleration–time pattern

In Chapter 2, we saw how important it is for a qualitative analyst to be able to interpret movement patterns such as displacement or angle time series. Force–time or acceleration–time patterns are far less likely to be encountered by the qualitative movement analyst and are not so revealing about other kinematic patterns – velocity and displacement. However, if you are prepared to accept that the velocity equals the area between the horizontal zero-acceleration line – the time (t) axis of the graph – and the acceleration curve from the start of the movement at time 0 up to any particular time, and that areas below the time axis are negative and those above positive, then several key points on the velocity–time graph follow. Ignore, for the time being, the numbers on the vertical axes of Figure 5.12.

- At time A in Figure 5.12(a), the area ($-A_1$) under the time axis and above the acceleration–time curve from 0 to A reaches its greatest negative value, so the vertical velocity of the jumper's centre of mass also reaches its greatest negative value there, corresponding to a zero acceleration. We can then sketch the velocity graph in Figure 5.12(b) up to time A.
- At time B in Figure 5.12(a), the area under the acceleration–time curve from A to B and above the time axis is $+A_1$ – the same magnitude as before but now positive. So the net area between the acceleration–time curve and the time axis from 0 to B is $-A_1 + A_1 = 0$. The vertical velocity of the jumper's centre of mass at B is, therefore, zero; we can now sketch the vertical velocity curve from A to B in Figure 5.12(b).
- From B to C the area between the acceleration–time curve and the time axis is positive, so the area (A_2) reaches its greatest positive value at C. Now we can sketch the vertical velocity curve from B to C in Figure 5.12(b).
- Finally, from C to take-off at D (time T), we have a small negative area, A_3, and the vertical velocity decreases from its largest positive value by this small amount before take-off, as in Figure 5.12(b).

Figure 5.12 Standing vertical jump time series: (a) acceleration; (b) velocity; (c) displacement.

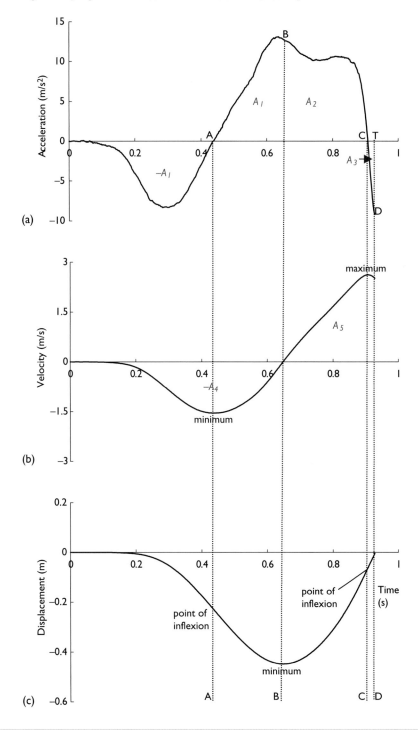

By similar reasoning, if you are prepared to accept that displacement of the jumper's centre of mass equals the area between the horizontal zero-velocity line – the time axis of the graph – and the velocity curve from the start of the movement at 0 up to any particular time, and, again, that areas below the time axis are negative and those above positive, then several key points on the displacement–time graph follow.

- At time B in Figure 5.12(b), the area ($-A_4$) under the time axis and above the vertical velocity–time curve from 0 to B reaches its greatest negative value, so the vertical displacement of the jumper's centre of mass also reaches its greatest negative value there, corresponding to a zero velocity. We can then sketch the displacement graph in Figure 5.12(c) up to B. Please note that this is the lowest point reached by the jumper's centre of mass at full hip and knee flexion, before the jumper starts to rise.
- At time T in Figure 5.12(b), the area (A_5) under the velocity–time curve from B to T and above the time axis is positive so the vertical displacement becomes less negative. Now we can sketch the vertical displacement curve from B to T in Figure 5.12(c). Note that the maximum displacement will occur after the person has left the force plate at the peak of the jump.

Note again, as in Chapter 2, the trend of peaks or, more obviously in this case, troughs is acceleration then velocity then displacement.

Quantitative evaluation of a force–time or acceleration–time pattern

Our volleyball coach may wish to obtain, from this acceleration–time curve, values for the magnitude of the vertical velocity at take-off and the maximum height reached by the player's centre of mass. The process of obtaining velocities and displacements from accelerations qualitatively was outlined in the previous section. The quantitative process for doing the same thing is referred to, mathematically, as integration; it can be performed graphically or numerically. If quantitative acceleration data are available – Figure 5.12(a) with the numbers shown on the axes if you like – we can integrate the acceleration data to obtain the velocity–time graph, as in Figure 5.12(b), which can, in turn, be integrated to give the displacement–time graph of Figure 5.12(c). Most quantitative ways of doing these integrations basically involve determining areas as above, but for increasing, small time intervals from left to right. A very slow but accurate way of doing this is counting areas under the curves drawn on graph paper. The resulting graphs, but with numbered axes as in Figure 5.12, would be more accurate than the 'sketched' qualitative ones but the shapes should be very similar if the qualitative analyst understood well the process of obtaining the velocity and displacement patterns. Our volleyball coach could now simply read the vertical take-off velocity (v_t) for each player from graphs such as Figure 5.12(b) at take-off, T (see Study task 2). We can calculate the maximum height (h) reached by a player's centre of mass by equating his or her take-off kinetic energy, $\frac{1}{2} m v_t^2$, to the potential energy at the peak of the jump, $m g h$, where m is the player's mass and g is gravitational acceleration, so $h = v_t^2/(2g)$.

DETERMINATION OF THE CENTRE OF MASS OF THE HUMAN BODY

We have seen in the previous sections that the most important application of the laws of linear motion in sports biomechanics is in expressing, completely, the motion of a sports performer's centre of mass, which is the unique point about which the mass of the performer is evenly distributed. The effects of external forces upon the sports performer can, therefore, be studied by the linear motion of the centre of mass and by rotations about the centre of mass. It is often found that the movement patterns of the centre of mass vary between highly skilled and less skilful performers, providing a simple tool for evaluating technique. Furthermore, the path of the centre of mass is important, for example, in studying whether a high jumper's centre of mass can pass below the bar while the jumper's body segments all pass over the bar, and how high above the hurdle a hurdler's centre of mass needs to be just to clear the hurdle.

The position of the centre of mass is a function of age, sex and body build and changes with breathing, ingestion of food and disposition of body fluids. It is doubtful whether it can be pinpointed to better than 3 mm. In the fundamental or anatomical reference position (see Chapter 1), the centre of mass lies about 56–57% of a male's height from the soles of the feet, the figure for females being 55%. In this position, the centre of mass is located about 40 mm inferior to the navel, roughly midway between the anterior and posterior skin surfaces. The position of the centre of mass is highly dependent on the orientation of a person's body segments. For example, in a piked body position the centre of mass of a gymnast may lie outside the body.

Historically, several techniques have been used to measure the position of the centre of mass of the sports performer. These included boards and scales and manikins (physical models). They are now rarely, if ever, used in sports bio-mechanics, having been superseded by the segmentation method. In this method, the following information is required to calculate the position of the whole body centre of mass:

- The masses of the individual body segments, usually as proportions of the total body mass or as regression equations.
- The locations of the centres of mass of those segments in the position to be analysed. This requirement is usually met by a combination of the pre-established location of each segment's centre of mass with respect to the end points of the segment, and the positions of those end points on a video image, for example the ankle joint in Figure 5.13(a). The end points can be joint centres or terminal points and are estimated in the analysed body position from the camera viewing direction (Figure 5.13). The anatomical landmarks used to estimate joint centre positions are shown in Table 5.1 and Box 6.2. It is worth noting in Figure 5.13(b) that even for this apparently sagittal plane movement in which the markers were carefully placed according to Table 5.1 and Box 6.2 with the lifter in the fundamental reference posture but with the palms facing backwards as when holding the bar, some markers no longer exactly overlay joint axes of rotation.

Figure 5.13 Determination of whole body centre of mass. (a) Simplified example for Study task 3; blue circles represent appropriate joint centres (for clarity, H = hip and W = wrist) or terminal points (see also Table 5.1). (b) Normal example with surface markers placed on the body in the reference position.

(a)

(b)

The ways of obtaining body segment data, including masses and locations of centres of mass, were briefly outlined in Chapter 4. The positions of the segmental end points are obtained from some visual record, usually a video recording of the movement. For the centre of mass to represent the system of segmental masses, the moment of mass (similar to the moment of force) of the centre of mass must be identical to the sum of the moments of body segment masses about any given axis. The calculation can be expressed symbolically as: $m\,r = \Sigma m_i\,r_i$, or $r = \Sigma(m_i/m)r_i$, where m_i is the mass of segment number i, m is the mass of the whole body (the sum of all of the individual segment masses Σm_i); (m_i/m is the fractional mass ratio of segment number i; and r and r_i are the position vectors, respectively, of the centre of mass of the whole body and the centre of mass of segment number i. In practice, the position vectors are specified by their two-dimensional (x, y) or three-dimensional (x, y, z) coordinates. Table 5.1 (see page 221) shows the calculation process for a two-dimensional case with given segmental mass fractions and position of centre of mass data (see also Study task 3).

FUNDAMENTALS OF ANGULAR KINETICS

Almost all human motion in sport and exercise involves rotation (angular motion), for example the movement of a body segment about its proximal joint. In Chapter 3, we considered the kinematics of angular motion. In this section, our focus will be on the kinetics of such motions.

Moments of inertia

In linear motion, the reluctance of an object to move (its inertia) is expressed by its mass. In angular motion, the reluctance of the object to rotate depends also on the distribution of that mass about the axis of rotation, and is expressed by the moment of inertia. The SI unit for moment of inertia is kilogram-metres^2 (kg m^2); moment of inertia is a scalar quantity. An extended (straight) gymnast has a greater moment of inertia than a piked or tucked gymnast and is, therefore, more 'reluctant' to rotate. This makes it more difficult to somersault in an extended, or layout, position than in a piked or tucked position. Formally stated, the moment of inertia is the measure of an object's resistance to accelerated angular motion about an axis. It is equal to the sum of the products of the masses of the object's elements and the squares of the distances of those elements from the axis of rotation. Moments of inertia can be expressed in terms of the radius of gyration, k, such that the moment of inertia (I) is the product of the mass (m) and the square of the radius of gyration: $I = m\,k^2$.

The moments of inertia for rotation about any point on a three-dimensional rigid body are normally expressed about three mutually perpendicular axes of symmetry; the moments of inertia about these axes are called the principal moments of inertia. For the sports performer in the anatomical reference position, the three principal axes of inertia through the centre of mass correspond with the three cardinal axes. The moment of inertia about the vertical axis is much smaller – about one-tenth – of the moment of inertia about the other two axes and that about the frontal axis is slightly less than that about the sagittal axis. It is worth noting that rotations about the intermediate principal axis, here the frontal axis, are unstable, whereas those about the principal axes with the greatest and smallest moments of inertia are stable. This is another factor making layout somersaults difficult, unless the body is realigned to make the moment of inertia about the frontal, or somersault, axis greater than that about the sagittal axis.

For the sports performer, movements of the limbs other than in symmetry away from the anatomical reference position result in a misalignment between the body's cardinal axes and the principal axes of inertia. This has important consequences for the generation of aerial twist in a twisting somersault.

BOX 5.3 LAWS OF ANGULAR MOTION

The laws of angular motion are analogous to Newton's three laws of linear motion.

Principle of conservation of angular momentum (law of inertia)

A rotating body will continue to turn about its axis of rotation with constant angular momentum unless an external torque (moment of force) acts on it. The magnitude of the torque about an axis of rotation is the product of the force and its moment arm, which is the perpendicular distance from the axis of rotation to the line of action of the force.

Law of momentum

The rate of change with time (d/dt) of angular momentum (L) of a body is proportional to the torque (M) causing it and has the same direction as the torque. This is expressed symbolically by the following equation, which holds true for rotation about any axis fixed in space: $M = dL/dt$. If we now rearrange this equation by multiplying by dt, and integrate it, we obtain: $\int M \, dt = \int dL = \Delta L$; that is, the impulse of the torque equals the change of angular momentum (ΔL). If the torque impulse is zero, this equation reduces to: $\Delta L = 0$ or L = constant, which is a mathematical statement of the first law of angular motion.

The last equation in the previous paragraph can be modified by writing $L = I\omega$ where I is the moment of inertia and ω is the angular velocity of the body if, and only if, the axis of rotation is fixed in space or is a principal axis of inertia of a rigid or quasi-rigid body. Then, and only then, $M = d(I\omega)/dt$. Furthermore if, and only if, I is constant (for example for an individual rigid or quasi-rigid body segment): $M = I\alpha$, where α is the angular acceleration. The restrictions on the use of the equations in this paragraph compared with the universality of the equations in the previous paragraph should be carefully noted.

The angular motion equations for three-dimensional rotation are far more complex than the ones above and will not be considered in this book.

The law of reaction

For every torque that is exerted by one object on another, an equal and opposite torque is exerted by the second object on the first. In Figure 5.14(a), the forward and upward swing of the long jumper's legs (thigh flexion) evoke a reaction causing the forward and downward motion of the trunk (trunk flexion). As the two torques are both within the jumper's body, there is no change in her angular momentum; the two torques involved produce equal but opposite impulses. In Figure 5.14(b), the ground provides the reaction to the action torque generated in the racket arm and

Figure 5.14 Action and reaction: (a) airborne; (b) with ground contact.

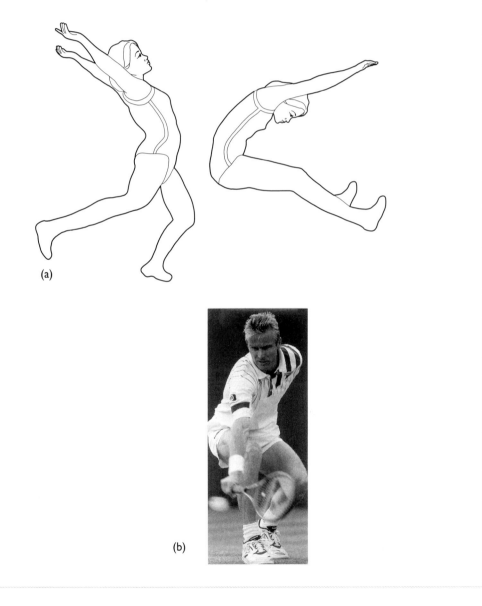

(a)

(b)

upper body, although this is not apparent because of the extremely high moment of inertia of the earth. If the tennis player played the same shot with his feet off the ground, there would be a torque on his lower body causing it to rotate counter to the movement of the upper body, lessening the angular momentum in the arm–racket system.

Minimising inertia

The law of momentum (Box 5.3) allows derivation of the principle of minimising inertia. The increase in angular velocity and, therefore, the reduction in the time taken to move through a specified angle, will be greater if the moment of inertia of the whole chain of body segments about the axis of rotation is minimised. Hence, for example, in running – and particularly in sprinting – the knee of the recovery leg is flexed to minimise the duration of the leg recovery phase.

Angular momentum of a rigid body

For a rigid body rotating about either an axis fixed in space, as in Figure 5.15(a), or a principal axis of inertia, it can then be shown that, for planar motion, the angular momentum (L) is the product of the body's moment of inertia about the axis (I) and its angular velocity (ω): that is, $L = I\omega$. The direction of the angular momentum vector L,

Figure 5.15 Angular momentum: (a) single rigid body; (b) part of a system of rigid bodies.

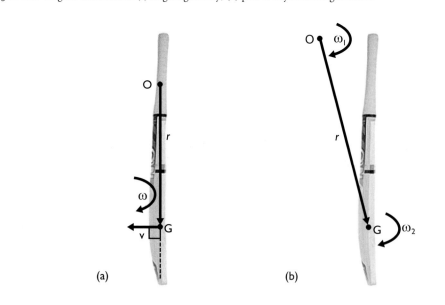

(a)

(b)

which is the same as that of $\boldsymbol{\omega}$, is given by allowing the position vector $\boldsymbol{r}$ to rotate towards the velocity vector $\boldsymbol{v}$ through the right angle indicated in Figure 5.15(a). By the right-hand rule, the angular momentum vector is into the plane of the page. The SI unit of angular momentum is kg m^2/s.

Angular momentum of a system of rigid bodies

For planar rotations of systems of rigid bodies, for example the sports performer, each rigid body can be considered to rotate about its centre of mass (G), with an angular velocity $\boldsymbol{\omega}_2$. This centre of mass rotates about the centre of mass of the whole system (O), with an angular velocity $\boldsymbol{\omega}_1$, as for the bat of Figure 5.15(b). The derivation will not be provided here, but the result is that the magnitude of the angular momentum of the bat is: $L = m\,r^2\,\omega_1 + I_g\,\omega_2$. The first term, owing to the motion of the body's centre of mass about the system's centre of mass, is known as the 'remote' angular momentum. The latter, owing to the rigid body's rotation about its own centre of mass, is the 'local' angular momentum. For an interconnected system of rigid body segments, representing the sports performer, the total angular momentum is the sum of the angular momentums of each of the segments calculated as in the above equation. The ways in which angular momentum is transferred between body segments can then be studied for sports activities such as airborne manoeuvres in gymnastics or the flight phase of the long jump.

GENERATION AND CONTROL OF ANGULAR MOMENTUM

A net external torque is needed to alter the angular momentum of a sports performer. Traditionally in sports biomechanics, three mechanisms of inducing rotation, or generating angular momentum, have been identified, although they are, in fact, related.

Force couple

A force couple consists of a parallel force system of two equal and opposite forces ($\boldsymbol{F}$) which are a certain distance apart (Figure 5.16(a)). The net translational effect of these two forces is zero and they cause only rotation. The net torque of the force couple is: $\boldsymbol{M} = 2\boldsymbol{r} \times \boldsymbol{F}$. The $\times$ sign in this equation tells us that the position vector, $\boldsymbol{r}$, and the force vector, $\boldsymbol{v}$, are multiplied vectorially (see Appendix 4.2). The resulting torque vector has a direction perpendicular to, and into, the plane of this page. Its magnitude ($2\,r\,F$) is the magnitude of one of the forces (F) multiplied by the perpendicular distance between them ($2r$).

The torque can be represented as in Figure 5.16(b) and has the same effect about a particular axis of rotation wherever it is applied along the body. In the absence of an

Figure 5.16 Generation of rotation: (a) force couple; (b) the torque (or moment) of the couple.

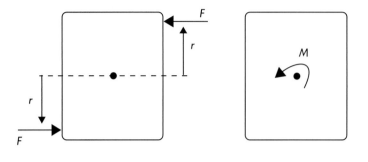

external axis of rotation, the body will rotate about an axis through its centre of mass. The swimmer in Figure 5.2(a) is acted upon by a force couple of her weight and the equal, but opposite, buoyancy force.

Eccentric force

An eccentric force ('eccentric' means 'off-centre') is effectively any force, or resultant of a force system, that is not zero and that does not act through the centre of mass of an object. This constitutes the commonest way of generating rotational motion, as in Figure 5.17(a).

The eccentric force here can be transformed by adding two equal and opposite forces at the centre of mass, as in Figure 5.17(b), which will have no net effect on the object. The two forces indicated in Figure 5.17(b) with an asterisk can then be considered together and constitute an anticlockwise force couple, which can be replaced by a torque M as in Figure 5.17(c). This leaves a 'pure' force (F) acting through the centre of mass, which causes only linear motion, $F = \mathrm{d}(m\ v)/\mathrm{d}t$; the torque M causes only rotation, $M = \mathrm{d}L/\mathrm{d}t$. The magnitude of the torque, M, is $F\ r$.

This example could be held to justify the use of the term 'torque' for the turning effect of an eccentric force although, strictly, 'torque' is defined as the moment of a force couple. The two terms, torque and moment, are often used interchangeably. There is a strong case for abandoning the use of the term moment of a force or couple entirely in favour of torque, given the various other uses of the term moment in biomechanics.

Checking of linear motion

Checking of linear motion occurs when an already moving body is suddenly stopped at one point. An example is the foot plant of a javelin thrower in the delivery stride, although the representation of such a system as a quasi-rigid body is of limited use.

Figure 5.17 Generation of rotation: (a) an eccentric force; (b) addition of two equal and opposite collinear forces acting through the centre of mass, G; (c) equivalence to a 'pure' force and a torque; (d) checking of linear motion.

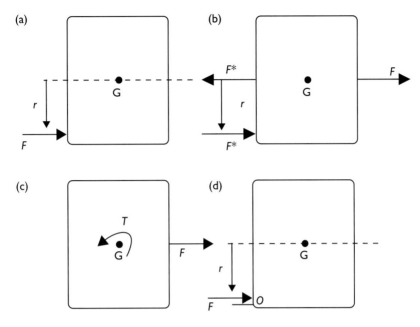

This, as shown in Figure 5.17(d), is merely a special case of an eccentric force. It is best considered in that way to avoid misunderstandings that exist in the literature, such as the misconception that O is the instantaneous centre of rotation. This last sentence begs the question of where the instantaneous centre of rotation does lie in such cases.

Consider a rigid body that is simultaneously rotating with angular velocity $\boldsymbol{\omega}$ about its centre of mass while moving linearly, as in Figure 5.18(a). The whole body has the same linear velocity ($\boldsymbol{v}$), as in Figure 5.18(b), but the tangential velocity owing to rotation ($\boldsymbol{v_t}$) depends on the displacement ($\boldsymbol{r}$) from the centre of mass, G, such that $\boldsymbol{v_t} = \boldsymbol{\omega} \times \boldsymbol{r}$, as in Figure 5.18(c). Adding $\boldsymbol{v}$ and $\boldsymbol{v_t}$ gives the net linear velocity (Figure 5.18(d)) that, at some point P (which need not lie within the body) is zero; this point is called the instantaneous centre of rotation and its position usually changes with time.

Consider now a similar rigid body acted upon by an impact force, as in Figure 5.18(e), which is similar to Figure 5.17(d). There will, in general, be a point R that will experience no net acceleration. By Newton's second law of linear motion, the magnitude of the acceleration of the centre of mass of the rigid body, mass m, is $a_g = F/m$. From Newton's second law of rotation (see above), the magnitude (F c) of the moment of the force $\boldsymbol{F}$ is equal to the product of the moment of inertia of the body about its centre of mass (I_g) and its angular acceleration (α). That is $F c = I_g \alpha$. This gives a tangential acceleration (α r) that increases linearly with distance (r) from G. The

Figure 5.18 Instantaneous centre of rotation and centre of percussion: (a) rigid body undergoing linear and angular motion; (b) linear velocity profile; (c) tangential velocity profile; (d) net velocity profile and instantaneous centre of rotation (P); (e) impact force on a rigid body at centre of percussion (Q); (f) net acceleration profile.

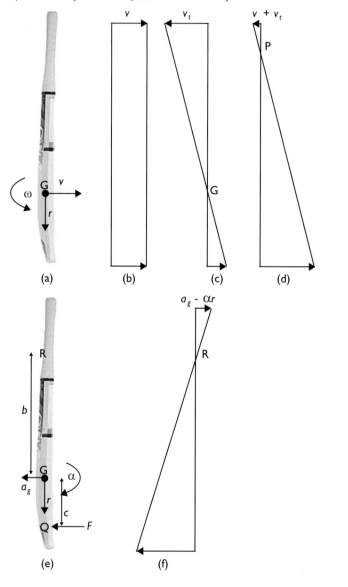

net acceleration profile then appears as in Figure 5.18(f). At point R ($r = b$), the two accelerations are equal and opposite and, therefore, cancel to give a zero net acceleration. The position of R is given by equating $F/m \, (= a_g)$ to $F \, b \, d/I_g$, the tangential acceleration at R. That is: $b \, c = I_g/m$. If R is a fixed centre of rotation then Q is known as the centre of percussion, defined as that point at which a force may be applied without causing an acceleration at another specific point, the centre of rotation. The above equation shows that Q and R can be reversed. The centre of percussion is important in sports in which objects, such as balls, are struck with other objects such as bats and rackets. If the impact occurs at the centre of percussion, no force is transmitted to the hands. For an object such as a cricket bat, the centre of percussion will lie some way below the centre of mass, whereas for a golf club, with the mass concentrated in the club head, the centres of percussion and mass more nearly coincide. Variation in grip position will alter the position of the centre of percussion. If the grip position is a long way from the centre of mass and the centre of percussion is close to the centre of mass, then the position of the centre of percussion will be less sensitive to changes in grip position. This is achieved by moving the centre of mass towards the centre of percussion, as for golf clubs with light shafts and heavy club heads, and cricket bats for which the mass of the bat is built up around the centre of percussion. Much tennis racket design has evolved towards positioning the centre of percussion nearer to the likely impact spot. The benefits of such a design feature include less fatigue and a reduction in injury.

The application of the centre of percussion concept to a generally non-rigid body, such as the human performer, is problematic. However, some insight can be gained into certain techniques. Consider a reversal of Figure 5.18(e) such that Q is high in the body and R is at the ground (similar to Figure 5.17(d)). Let F be the reaction force experienced by a thrower who is applying force to an external object. If F is directed through the centre of percussion, there will be no resultant acceleration at R (the foot–ground interface). A second example relates Figure 5.18(e) to the braking effect when the foot lands in front of the body's centre of mass. The horizontal component of the impact force will oppose relative motion and cause an acceleration distribution, as in Figure 5.18(f), with all body parts below R decelerated and only those above R accelerated. This is important in, for example, javelin throwing, where it is desirable not to slow the speed of the object to be thrown during the final foot contacts of the thrower.

Transfer of angular momentum

The principle of transfer of angular momentum from segment to segment is sometimes considered to be a basic principle of coordinated movement. Consider, for example, the skater in Figure 4.11; if she moved her arms fully away from her, to a 90° abducted position, she would decrease her speed of rotation; if she moved them into her body, she would rotate faster. This is sometimes interpreted in terms only of the two 'quasi-static' end positions – fully abducted arms, and arms drawn in to the body. The former position has a large moment of inertia, and hence low speed of rotation; the latter position has a low moment of inertia and high speed of rotation. However, from the abducted-arms to the tucked-arms position, the arms lose angular momentum as they move towards the body, 'transferring' some of their angular momentum to the rest of the body which, therefore, turns faster. A more complex example is the hitch kick technique in the flight phase of the long jump, in which the arm and leg motions transfer angular momentum from the trunk to the limbs to prevent the jumper from rotating forwards too early in the flight phase.

Trading of angular momentum

The term 'trading' of angular momentum is often used to refer to the transfer of angular momentum from one axis of rotation to another. For example, the model diver – or gymnast – in Figure 5.19 takes off with angular momentum ($L = L_{som}$) about the somersault, or horizontal, axis. The diver then adducts her left arm, or performs some other asymmetrical movement, by a muscular torque that evokes an equal but opposite counter-rotation of the rest of the body to produce an angle of tilt. No external torque has been applied so the angular momentum (L) is still constant about a horizontal axis but now has a component (L_{twist}) about the twisting axis. The diver has 'traded' some somersaulting angular momentum for twisting, or longitudinal, angular momentum and will now both somersault and twist. It is often argued that this method of generating twisting angular momentum is preferable to 'contact twist' (twist generated when in contact with an external surface), as it can be more easily removed by re-establishing the original body position before landing. This can avoid problems in gymnastics, trampolining and diving caused by landing with residual twisting angular momentum. The crucial factor in generating airborne twist is to establish a tilt angle and, approximately, the twist rate is proportional to the angle of tilt. In practice, many sports performers use both the contact and the airborne mechanisms to acquire twist.

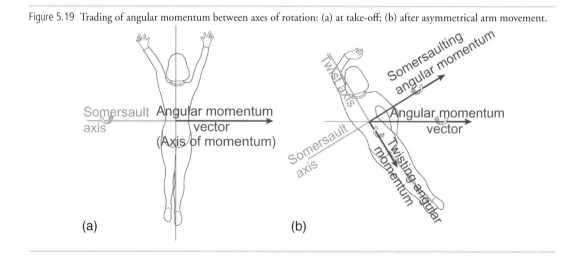

Figure 5.19 Trading of angular momentum between axes of rotation: (a) at take-off; (b) after asymmetrical arm movement.

Three-dimensional rotation

Rotational movements in airborne activities in diving, gymnastics and trampolining, for example, usually involve three-dimensional, multi-segmental movements. The mathematical analysis of the dynamics of such movements is beyond the scope of this book. The three-dimensional dynamics of even a rigid or quasi-rigid body are far from simple. For example, in two-dimensional rotation of such bodies, the angular momentum and angular velocity vectors coincide in direction. If a body, with principal moments of inertia that are not identical – as is the case for the sports performer – rotates about an axis that does not coincide with one of the principal axes, then the angular velocity vector and the angular momentum vectors do not coincide. A movement known as 'nutation' can result. Nutation also occurs, for example, when performing an airborne pirouette with asymmetrical arm positions. The body's longitudinal axis is displaced away from its original position of coincidence with the angular momentum vector, sometimes called the axis of momentum, and will describe a cone around that vector. Furthermore, the equation of conservation of angular momentum applies to an inertial frame of reference, such as one moving with the centre of mass of the performer but always parallel to a fixed, stationary frame of reference. The conservation of angular momentum does not generally apply to a frame of reference fixed in the performer's body and rotating with it.

MEASUREMENT OF FORCE

BOX 5.4 WHY MEASURE FORCE OR PRESSURE?

- To provide further movement patterns for the use of the qualitative analyst.
- To highlight potential risk factors, particularly in high-impact activities.
- To evaluate, for example, the foot-strike patterns of runners or the balance of archers.
- To provide the external inputs for internal joint moment and force calculations (inverse dynamics; see Bartlett, 1999; Further Reading, page 222).
- The forces and pressures can be further processed to provide other movement information (see below).

Most force measurements in sport use a force plate, which measures the contact force components (Figure 5.20) between the ground, called the ground contact force, or another surface, and the sports performer. The measured force acting on the performer has the same magnitude as, but opposite direction from, the reaction force exerted on the performer by the force plate, by the law of action–reaction.

Force plates are widely used in research into the loading on the various joints of the body. In such research, movement data from videography, or another motion analysis system, are used with force, torque and centre of pressure data to calculate the resultant forces and torques at body joints. The two recording systems need to be synchronised for such investigations.

Force plates can be obtained in a variety of sizes. The most commonly used have a relatively small contact area, for example 600 × 400 mm for the Kistler type 9281B11 (Kistler Instrument Corporation, Winterthur, Switzerland; http://www.kistler.com) or 508 × 643 mm for the AMTI model 0R6–5–1 (Advanced Mechanical Technology Incorporated, Watertown, MA, USA; http://www.amtiweb.com) and weigh between 310 and 410 N, although much lighter plates have recently become available. They are normally bolted to a base plate set in concrete.

Forces of interaction between the sports performer and items of sports equipment can also be measured using other force transducers, usually purpose-built or adapted for a particular application. Such transducers have been used, for example, to measure the forces exerted by a rower on an oar, or by a cyclist on his or her bike's pedals. Many of the principles discussed in this chapter for force plates also apply to force transducers in general. Further consideration of the operation of these, usually specialist, devices will not be undertaken in this book.

One limitation of force plates is that they do not show how the applied force is distributed over the contact surface, for example the shoe or the foot. This information can be obtained from pressure plates, pads and insoles, which will be considered in the next section.

Force plates used to evaluate sporting performance are sophisticated electronic devices and are generally very accurate. Essentially, they can be considered as weighing

Figure 5.20 Ground contact force (F_x, F_y, F_z) and moment (or torque) (M_x, M_y, M_z) components that act on the sports performer.

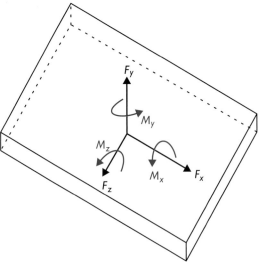

systems that are responsive to changes in the displacement of a sensor or detecting element. They incorporate a force transducer, which converts the force into an electrical signal. The transducers are mounted on the supports of the rigid surface of the force plate, usually one support at each of the four corners of a rectangle. One transducer is used, at each support, to measure each of the three force components, one of which is perpendicular to the plate and two tangential to it; for ground contact forces, there are usually one vertical and two horizontal force components, and we will use this example in what follows. The transducers are normally strain gauges, as in the AMTI plates, or piezoelectric, as in the Kistler plates. The signals from the transducers are amplified and may undergo other electrical modification. The amplified and modified signals are converted to digital signals for computer processing. The signal is then sampled at discrete time intervals, expressed as the sampling rate or sampling frequency. The Nyquist sampling theorem (see Chapter 4) requires a sampling frequency at least twice that of the highest signal frequency. It should be remembered that, although the frequency content of much human movement is low, many force plate applications involve impacts, which have a higher frequency content. A sampling frequency as high as 500 Hz or 1 kHz may, therefore, be appropriate.

Accurate (valid) and reliable force plate measurements depend on adequate system sensitivity, a low force detection threshold, high linearity, low hysteresis, low crosstalk and the elimination of cable interference, electrical inductance and temperature and humidity variations. The plate must be sufficiently large to accommodate the movement under investigation. Extraneous vibrations must be excluded. Mounting instructions for force plates are specified by the manufacturers. The plate is normally sited on the bottom floor of a building in a large concrete block. If mounted outdoors, a large concrete block sited on pebbles or gravel is usually a suitable base and attention must be given to problems of drainage. The main measurement characteristics of a force plate are considered below. In addition, a good temperature range (−20 to 70°C) and a relatively light weight may be important. Also, any variation in the recorded force with the position on the plate surface at which it is applied should be less than 2–3% in the worst case.

Force plate characteristics

Linearity

Linearity is expressed as the maximum deviation from linearity as a percentage of full-scale deflection. For example, in Figure 5.21(a), linearity would be expressed as $y/Y \times 100\%$. Although good linearity is not essential for accurate measurements, as a non-linear system can be calibrated, it is useful and does make calibration easier. A suitable figure for a force plate for use in sports biomechanics would be 0.5% of full-scale deflection or better.

Figure 5.21 Force plate characteristics: (a) linearity; (b) hysteresis (simplified); (c) output saturation and effect of range on sensitivity.

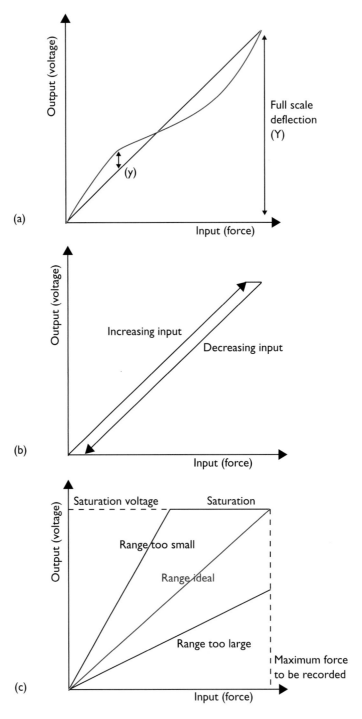

Hysteresis

Hysteresis exists when the input–output relationship depends on whether the input force is increasing or decreasing, as in Figure 5.21(b). Hysteresis can be caused, for example, by the presence of deforming mechanical elements in the force transducers. It is expressed as the maximum difference between the output voltage for the same force, increasing and decreasing, divided by the full-scale output voltage. It should be 0.5% of full-scale deflection or less.

Range

The range of forces that can be measured must be adequate for the application and the range should be adjustable. If the range is too small for the forces being measured, the output voltage will saturate (reach a constant value) as shown in Figure 5.21(c). Suitable maximum ranges for many sports biomechanics applications would be −10 to +10 kN for the two horizontal axes and −10 to +20 kN for the vertical axis.

Sensitivity

Sensitivity is the change in the recorded signal for a unit change in the force input, or the slope of the idealised linear voltage–force relationships of Figure 5.21. The sensitivity decreases with increasing range. Good sensitivity is essential as it is a limiting factor on the accuracy of the measurement. In most modern force plate systems, an analog-to-digital (A–D) converter is used and is usually the main limitation on the overall system resolution. An 8-bit A–D converter divides the input into an output that can take one of 256 (2^8) discrete values. The resolution is then 100/255%, approximately 0.4%. A 12-bit A–D converter will improve the resolution to about 0.025%. In a force plate system with adjustable range, it is essential to choose the range that just avoids saturation, so as to achieve optimum sensitivity. For example, consider that the maximum vertical force to be recorded is 4300 N, an 8-bit A–D converter is used, and a choice of ranges of 10 000 N, 5000 N and 2500 N is available. The best available range is 5000 N, and the force would be recorded approximately to the nearest 20 N (5000/255). The percentage error in the maximum force is then only about 0.5% (20 × 100/4300). A range of 10 000 N would double this error to about 1%. A range of 2500 N would cause saturation, and the maximum force recorded would be 2500 N, an underestimate of over 40% (compare with Figure 5.21(c)).

Crosstalk

Force plates are normally used to measure force components in more than one direction. The possibility then exists of forces in one component direction affecting the forces recorded by the transducers used for the other components. The term crosstalk is used to express this interference between the recording channels for the various force components. Crosstalk must be small, preferably less than 3% of full-scale deflection.

Dynamic response

Forces in sport almost always change rapidly as a function of time. The way in which the measuring system responds to such rapidly changing forces is crucial to the accuracy of the measurements and is called the 'dynamic response' of the system. The considerations here relate mostly to the mechanical components of the system. A representation of simple sinusoidally varying input and output signals of a single frequency (ω) as a function of time is presented in Figure 5.22.

The ratio of the amplitudes (maximum values) of the output to the input signal is called the amplitude ratio (A). The time by which the output signal lags the input signal is called the time lag. This is often expressed as the phase angle or phase lag (ϕ), which is the time lag multiplied by the signal frequency.

In practice, the force signal will contain a range of frequency components, each of which could have different phase lags and amplitude ratios. The more different these are across the range of frequencies present in the signal, the greater will be the distortion of the output signal. This will reduce the accuracy of the measurement. We therefore require the following:

- All the frequencies present in the force signal should be equally amplified; this means that there should be a constant value of the amplitude ratio A (system calibration can allow A to be considered as 1, as in Figure 5.23).
- The phase lag (ϕ) should be small.

Figure 5.22 Representation of force input and recorded output signals as a function of time.

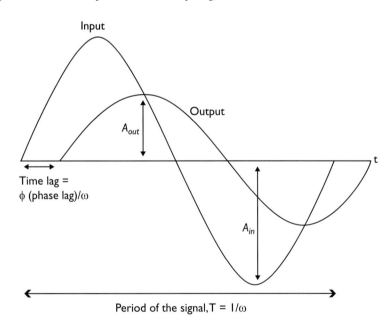

A force plate is a measuring system consisting essentially of a mass (m), a spring of stiffness k, and a damping element (c), rather like a racing car's suspension system; some simple models of the human lower extremity also use such a mass–spring–damper model. The 'steady-state' frequency response characteristics of such a system are usually represented by the unique series of non-dimensional curves of Figure 5.23. The response of such a system to an instantaneous change of the input force is known as its 'transient' response and can be represented as in Figure 5.24.

In these two figures:

- The 'damping ratio' of the system, $\zeta = c/\sqrt{(4\ k\ m)}$
- The 'frequency ratio' (ω/ω_n) is the ratio of the signal frequency (ω) to the 'natural frequency (ω_n)' of the force plate
- The natural frequency $\omega_n = \sqrt{(k/m)}$, is the frequency at which the force plate will vibrate if struck and then allowed to vibrate freely.

Figure 5.23 Steady-state frequency response characteristics of a typical second-order force plate system: (a) amplitude plot; (b) phase plot for damping ratios of: 2 (overdamped case shown by the dashed blue curve), 0.707 (critically damped case shown by the continuous black curve), and 0.2 (underdamped case shown by the continuous blue curve).

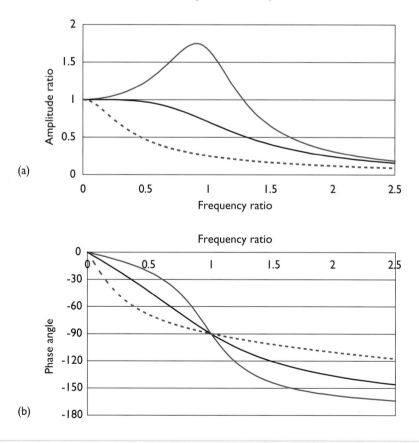

Figure 5.24 Transient response characteristics of a typical second-order force plate system for damping ratios of: 2 (over-damped case shown by the dashed blue curve), 0.707 (critically damped case shown by the continuous black curve), and 0.2 (underdamped case shown by the continuous blue curve). The output signal 'settles' to the input signal (the dashed black horizontal line) far quicker for the critically damped case than for the other two.

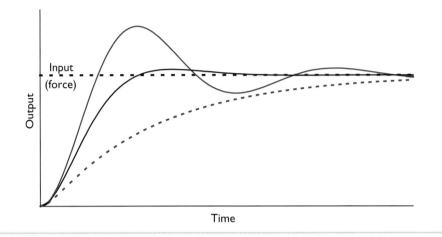

To obtain a suitable transient response, the damping ratio needs to be around 0.5–0.8 to avoid under- or over-damping. A value of 0.707 is considered ideal, as in Figure 5.24, as this ensures that the output follows the input signal – reaches steady state – in the shortest possible time.

Careful inspection of Figure 5.23 allows the suitable frequency range to be established for the steady-state response to achieve a nearly constant amplitude ratio (Figure 5.23(a)) and a small phase lag (Figure 5.23(b)), with $\zeta = 0.707$. The frequency ratio ω'/ω_n, where ω' is the largest frequency of interest in the signal, must be small, ideally less than 0.2. Above this value, the amplitude ratio (Figure 5.23(a)) starts to deviate from a constant value and the phase lag increases and becomes very dependent on the frequency, causing errors in the output signal. For impact forces, the natural frequency should be 10 times the equivalent frequency of the impact, which can exceed

BOX 5.5 GUIDELINE VALUES FOR FORCE PLATE CHARACTERISTICS

Linearity	≤0.5% of full-scale deflection
Hysteresis	≤0.5% of full-scale deflection
Crosstalk	≤3.0%
Natural frequency	≥800 Hz
Maximum frequency ratio (of signal frequency to natural frequency)	≤0.2
Damping ratio	0.5 to 0.8 (0.707 optimum)

100 Hz in sports activities. The natural frequency must, therefore, be as large as possible to record the frequencies of interest. The structure must be relatively light but stiff, to give a high natural frequency; this consideration relates not only to the plate but also to its mounting. A high natural frequency, of around 1000 Hz, would be desirable for most applications in sports biomechanics. Among the highest natural frequencies specified for commercial plates are: for one piezoelectric plate 850 Hz, all three channels; for two strain gauge plates (a) 1000 Hz for the vertical channel, 550 Hz for the two horizontal channels; and (b) 1500 Hz (vertical) and 320 Hz (horizontal). However, the value specified for a particular plate may not always be found in practical applications.

Experimental procedures

General

A force plate, if correctly installed and mounted and used with appropriate auxiliary equipment, is generally simple to use. When used with an A–D converter and computer, the timing of data collection is important. This can be achieved, for example, by computer control of the collection time, by the use of photoelectric triggers or by sampling only when the force exceeds a certain threshold. When using a force plate with video recording, synchronisation of the two will be required. This can be done in several ways, such as causing the triggering of the force plate data collection to illuminate a light in the field of view of the cameras (see also Study task 5).

The sensitivity of the overall system will need to be adjusted to prevent saturation while ensuring the largest possible use of the equipment's range. This can often be done by trial-and-error adjustments of amplifier gains; this should obviously be done before the main data collection. If the manufacturers have recommended warm-up times for the system amplifiers, then these must be carefully observed.

Care must be taken in the experimental protocol, if the performer moves on to the plate, to ensure that foot contact occurs with little (preferably no), targeting of the plate by the performer. This may, for example, require the plate to be concealed by covering the surface with a material similar to that of the surroundings of the plate. Obviously the external validity of an investigation will be compromised if changes in movement patterns occur to ensure foot strike on the plate. Also important for external – ecological – validity, particularly when recording impacts, is matching the plate surface to that which normally exists in the sport being studied. The aluminium surfaces normally used for force plates are unrepresentative of most sports surfaces.

Calibration

It is essential that the force plate system can be calibrated to minimise systematic errors. The overall system will require regular calibration checks, even if this is not necessary for the force plate itself. Calibration of the amplifier output as a function of force input will usually be set by the manufacturers and may require periodic checking. The vertical channel is easily calibrated under static loading conditions by use of known weights. If these are applied at different points across the plate surface, the variability of the recorded force with its point of application can also be checked. The horizontal channels can also be statically calibrated, although not so easily. One method of doing this involves attaching a cable to the plate surface, passing the cable over a frictionless air pulley at the level of the plate surface, and adding weights to the free end of the cable. Obviously this cannot be done while the plate is installed in the ground flush with the surrounding surface. There appears to be little guidance provided to users on the need for, or regularity of, dynamic calibration checks on force plates. The tendency of piezoelectric transducers to drift may mean that zero corrections are required and strain gauge plates may need more frequent calibration checks than do piezoelectric ones.

Crosstalk can be checked by recording the outputs from the two horizontal channels when only a vertical force, such as a weight, is applied to the plate. A similar procedure can be used for assessing crosstalk on the vertical channel if horizontal forces can be applied. Positions of the point of force application can be checked by placing weights on the plate at various positions and comparing these with centre of pressure positions calculated from the outputs from the individual vertical force transducers. As errors in these calculations are problematic when small forces are being recorded, small as well as large weights should be included in such checks. Finally, the natural frequency can be checked by lightly striking the plate with a metal object and using an oscilloscope to show the ringing of the plate at its natural frequency. This should be carried out, of course, in the location in which the plate is to be used.

Data processing

Processing of force plate signals is relatively simple and accurate, compared with most data in sports biomechanics. The example data of Figure 5.25 were obtained from a standing broad (long) jump. The three mutually perpendicular (orthogonal) components of the ground contact force (Figures 5.20 and 5.25(a)) are easily obtained by summing the outputs of individual transducers. As the plate provides whole body measurements, these forces (F) can be easily converted to the three components of centre of mass acceleration (a) simply by dividing by the mass of the performer ($F = m\,a$, Figure 5.25(b)), after subtracting the performer's weight from the vertical force component. The coordinates of the point of application of the force, the centre of pressure (Figure 5.25(c)), on the plate working surface can also be calculated. The accuracy of the centre of pressure calculations in particular depends on careful calibration of the force plate; this accuracy deteriorates at the beginning and end of any contact phase, when the calculation of centre of pressure involves the division of small forces by other small forces.

The moment of the ground contact force about the vertical axis perpendicular to the

Figure 5.25 Force plate variables (as sagittal [blue], frontal [dashed blue] and vertical [black] components except for power) as functions of time for a standing broad jump: (a) force; (b) centre of mass acceleration; (c) point of force application (x, z only); (d) moment; (e) centre of mass velocity; (f) whole body angular momentum; (g) centre of mass position; (h) load rate; (i) whole body power.

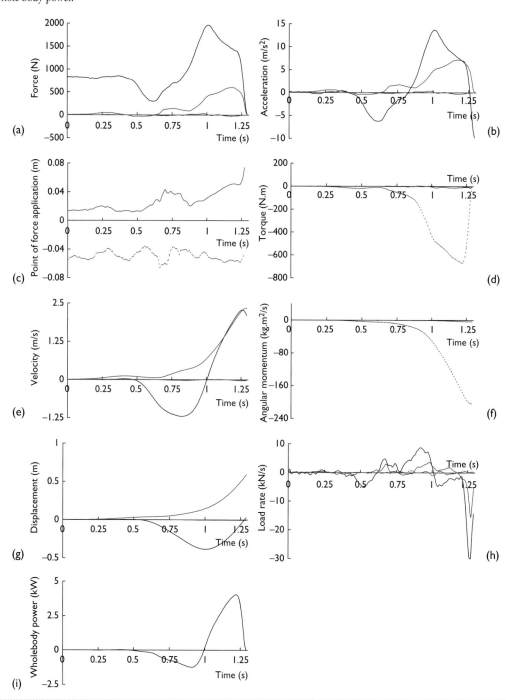

plane of the plate – the frictional torque, sometimes called the free moment (Figure 5.25(d)) – can be easily calculated. With appropriate knowledge of the position of the performer's centre of mass, in principle at one instant only, the component torques (moments) of the ground contact force (Figure 5.25(d)) can also be calculated about the two horizontal and mutually perpendicular axes passing through the performer's centre of mass and parallel to the plate.

From the force–time and torque–time data, integration can be performed to find overall or instant-by-instant changes in centre of mass velocity and whole-body angular momentum. Absolute magnitudes of these variables at all instants (Figures 5.25(e) and (f)) can be calculated only if their values are known at least at one instant. These values could be obtained from videography or another motion analysis system. They are easily obtained if the performer is at rest on the plate at some instant, when both linear and angular momentums are zero. If absolute velocities are known, then the changes in position – displacement – can be found by integration. In this case, absolute values of the position of the centre of mass with respect to the plate coordinate system (Figure 5.25(g)) can be obtained if that position is known for at least one instant. Again, that value could be obtained from videography or another motion analysis system. Alternatively, the horizontal coordinates can be obtained from the centre of pressure position with the person stationary on the plate, and the vertical coordinate as a fraction of the person's height.

Figure 5.26 (a) Side view of force vectors for a standing broad jump; (b) centre of pressure path from above. (S shows the start of the movement and T take-off.) Note: (a) and (b) are not to the same scale.

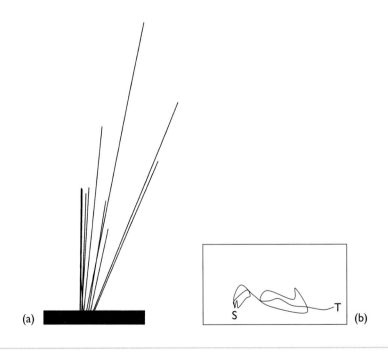

The load rate (Figure 5.25(h)) can be calculated as the rate of change with time t (d/dt) of the contact force F (dF/dt). The load rate has often been linked to injury. Other calculations that can be performed include whole body power ($P = F.v$) (Figure 5.25(i)), which does not have x, y and z components, as power is a scalar (see also Appendix 4.2).

All the above variables can be presented graphically as functions of time, as in Figure 5.25, providing the qualitative analyst with a rich new set of movement patterns and the quantitative analyst with further useful data. In addition, the forces acting on the performer can be represented as instantaneous force vectors, arising from the instantaneous centres of pressure (a side view is shown in Figure 5.26(a)). Front, top and three-dimensional views of the force vectors are also possible. The centre of pressure path can also be shown superimposed on the plate surface (Figure 5.26(b)).

MEASUREMENT OF PRESSURE

As we noted in the previous section, force plates provide the position of the point of application of the force, also called the centre of pressure, on the plate. This is the point at which the force can be considered to act, although the pressure is distributed over the plate and foot. Indeed, there may be no pressure acting at the centre of pressure when, for example, it is below the arch of the foot or between the feet during double stance. Information about the distribution of pressure over the contacting surface would be required, for example, to examine the areas of the foot on which forces are concentrated during the stance phase in running to improve running shoe design. In such cases, a pressure platform or pressure pad must be used. These devices consist of a set of force transducers with a small contact area over which the mean pressure for that area of contact (pressure = force ÷ area) is calculated. For just a few selected regions of the contact surface, pressures can be measured using individual sensors; problems with this approach include choosing the appropriate locations, and movement of the sensors during the activity being studied.

Various types of pressure plate are commercially available, and have been mostly used for the measurement of pressure distributions between the foot or shoe and the ground. They have been used only to a limited extent in sport, partly because their usually small size causes targeting problems for the performer. Also, the distribution of pressure is altered if the platform is covered with a sports surface. Pressure pads have been developed for the measurement of contact pressures between parts of the body and the surroundings. Specialist applications have included pads for the study of the dorsal pressures on the foot within a shoe. Plantar pressure insoles are commercially available (for example, Figure 5.27) and can be used to measure the plantar pressure distribution between the foot and the shoe. This is generally more important, for the sports performer, than the pressure between the shoe and the ground measured by pressure plates. These insoles allow data collection for several foot strikes and do not cause problems of targeting.

Figure 5.27 A plantar pressure insole system – Pedar™ (Novel GmbH, Munich, Germany; www.novel.de).

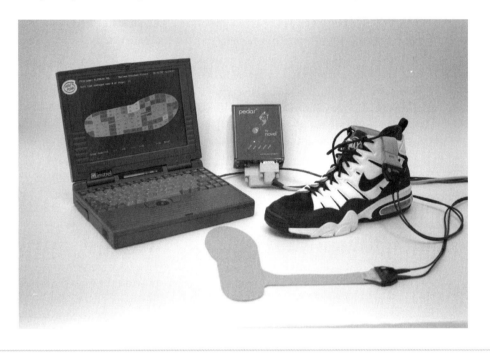

All pressure insoles have certain drawbacks for use in sport. First, they require cabling and a battery pack to be worn by the performer. Secondly, the insole may alter the pressure distribution because of its thickness. Thirdly, all commercially available devices currently only record the normal component of stress (pressure), not the tangential (shear) components. Furthermore they are susceptible to mechanical damage and cross-talk between the individual sensors. The durability of the sensors is a function of thickness – the thicker they are, the more durable – but mechanical crosstalk also increases with sensor thickness. Although very thin sensors may be suitable for static and slowly changing pressure measurements, durability is important in sports applications involving rapid pressure changes, such as foot strike. Pressure plates, pads and insoles suitable for use in sports biomechanics are based on capacitive, conductive or piezoelectric transducers (for further details, see Lees and Lake, 2007; Further Reading, page 222).

Data processing

Many data processing and data presentation options are available to help analyse the results from pressure measuring devices. In principle, all of the data processing options available for force plates also apply to pressure plates. In addition, displays of whole foot pressure distributions are available, similar to those below for pressure insoles. The centre of pressure path – the gait line – measured from pressure insoles and plates is far less sensitive to error than for force plates. Pressure insole software usually provides for many displays of the pressure–time information. The software also allows the user to define areas of the foot of special interest (up to eight normally) and then to study those areas individually as well as the foot as a whole. The information available includes the following:

- Three-dimensional 'colour'-coded wire frame displays of the summary maximum pressure distributions at each sensor (the maximum pressure picture) and the pressure distribution at each sample time (a monochrome version of which is shown in Figures 5.28(a) and (b)), displayed as if viewing the right foot from below or the left from above; these can usually be animated so that the system user can see how the pressure patterns evolve over time. These are fascinating patterns of movement that, to date, have been little used by qualitative analysts.
- Force (sum of all sensor readings), maximum pressure and contact area as functions of time for the whole foot and any specified region of the foot.
- Bar charts of the peak pressure, force, contact area and the pressure–time integral for all regions of the foot.
- The centre of pressure path, or gait line (Figure 5.28(c)).

SUMMARY

In this chapter we considered linear 'kinetics', which is important for an understanding of human movement in sport and exercise. This included the definition of force, the identification of the various external forces acting in sport and how they combine, and the laws of linear kinetics and related concepts, such as linear momentum. We addressed how friction and traction influence movements in sport and exercise, including reducing and increasing friction and traction. Fluid dynamic forces were also considered; the importance of lift and drag forces on both the performer and on objects for which the fluid dynamics can impact on a player's movements were outlined. We emphasised both qualitative and quantitative aspects of force–time graphs. The segmentation method for calculating the position of the whole body centre of mass of the sports performer was explained. The vitally important topic of rotational kinetics was covered, including the laws of rotational kinetics and related concepts such as angular momentum and the ways in which rotation is acquired and controlled in sports motions. The use of force plates in sports biomechanics was covered, including the equipment and methods used, and the processing of force plate data. We also considered the important measurement characteristics required for a force plate in sports biomechanics. The procedures for calibrating a force plate were outlined, along with those used to record forces in practice. The different ways in which force plate data can be processed to obtain other movement variables were covered. The value of contact pressure measurements in the study of sports movements was covered. Some examples were provided of the ways in which pressure transducer data can be presented to aid analysis of sports movements.

Figure 5.28 Pedar™ insole data displays: (a, b) pressure distributions as three-dimensional wire frame displays of (a) summary maximum pressures (maximum pressure picture); (b) pressure distribution at one sample time; (c) centre of pressure paths (F indicates first and L last contact).

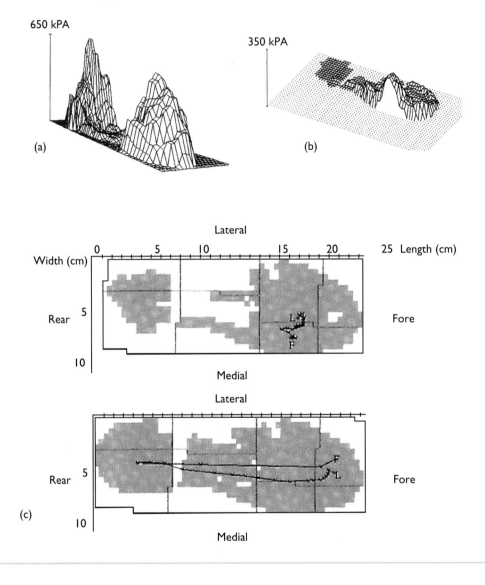

STUDY TASKS

1 (a) List the external forces that act on the sports performer and, for each force, give an example of a sport or exercise in which that force will be very important.

(b) Define and explain the three laws of linear kinetics and give at least two examples from sport or exercise, other than the examples in this chapter, of the application of each law.

Hint: You may wish to reread the sections on 'Forces in sport' (pages 164–80) and 'Momentum and the laws of linear motion' (pages 183–5) before undertaking this task.

2   Download a force–time Excel spreadsheet for a standing vertical jump from the book's website. The sample times (t) and magnitudes of the vertical ground reaction forces (F) are shown in the first two columns of the spreadsheet.

   (a) Obtain the vertical accelerations (a) in the third column by noting the jumper's weight (G) when standing still at the start of the sequence and using $a = g\, F/G$, where $g = 9.81$ m/s^2.

   (b) Use a simplified numerical integration formula for the change in the magnitude of the vertical velocity from one time interval, i, to the next, i+1, over sampling time Δt: $\Delta v = (a_i + a_{i+1})\Delta t/2$, and noting $v = 0$ at $t = 0$. Put your velocities in the fourth column of the spreadsheet.

   (c) Using a similar numerical integration formula for the change in magnitude of the centre of mass vertical displacement (y) from one time interval, i, to the next, i+1, over sampling time Δt: $\Delta y = (v_i + v_{i+1})\Delta t/2$, and defining $y = 0$ at $t = 0$. Put your displacements in the fifth column of the spreadsheet.

   (d) Plot the time series of vertical force, and centre of mass vertical acceleration, velocity and displacement. Compare your answers with Figure 5.12.

   (e) What was the jumper's take-off velocity and what was the maximum height reached by the centre of mass?

   Hint: You should reread the section on 'Force–time graphs as movement patterns' (pages 186–8) before undertaking this task. If you are unfamiliar with performing simple calculations in Microsoft Excel, go to their online help site, or see your tutor.

3   Photocopy Figure 5.13(a) or download it from the book's website. Measure the x and y coordinates of each of the segment end points. Then use a photocopy of Table 5.1, or download it from the website, to calculate the position of the ski jumper's whole body centre of mass in the units of the image. Assume that the joints on the right side of the body have identical coordinates to those on the left side. Finally, as a check on your calculation, mark the resulting centre of mass position on your figure. If it looks silly, check your calculations and repeat until the centre of mass position appears reasonable. Then repeat for Figure 5.13(b).

   Hint: You may wish to reread the section on 'Determination of the centre of mass of the human body' (pages 189–91) before undertaking this task.

4   Carry out the inclined plane experiment, mentioned on pages 167–8, to calculate the coefficient of friction between the material of a sports surface and a training shoe and other sports objects. You only need a board covered with relevant material, a shoe and a protractor.

   Hint: You may wish to reread the subsection on 'Friction' (pages 166–7) before undertaking this task.

5   Obtain a video recording of top-class diving, trampolining or gymnastics from your university resources or from a suitable website. Carefully analyse some airborne

movements that do not involve twisting, including the transfer of angular momentum between body segments. Repeat for movements involving twisting; consider in particular the ways in which the performers generate twist in somersaulting movements.

Hint: You should reread the section on 'Generation and control of angular momentum' (pages 195–201) before undertaking this task.

6 (a) Outline how you would statically calibrate a force plate, how you would check for variability of the recorded force with its point of application on the plate surface, how you would check for crosstalk, and how you would check the accuracy of centre of pressure calculations.

(b) Describe two ways in which you might be able to synchronise the recording of forces from a force plate with a video recording of the movement. This may require some careful thought.

Hint: You may wish to reread the subsection on 'Experimental procedures' in 'Measurement of force' (pages 209–10) before undertaking this task.

7 If you have access to a force plate, perform an experiment involving a standing broad (long) jump from the plate, with arm countermovements. [If you do not have access to a force plate, you will find an Excel spreadsheet containing force plate data from a broad jump on this book's website. From the recorded force components, see if you can obtain the other data that were covered in the subsection on 'Data processing' in 'Measurement of force' (pages 210–13). If your force plate software supports all the processing options for these data, perform these calculations for all three force channels. Compare the results you obtain with those of Figure 5.25.

Hint: You should reread the subsection on 'Data processing' (pages 210–13) and revisit Study task 2 before undertaking this task.

8 Visit the book's website and look at the various examples there – from pressure plates and insoles – of the movement patterns available to a qualitative movement analyst. Assess which of these you think could be of routine use to a qualitative analyst and which are best thought of as back-up information that might occasionally be useful.

Hint: You might wish to reread relevant sections on movement patterns in Chapter 3 before undertaking this task.

You should also answer the multiple choice questions for Chapter 5 on the book's website.

GLOSSARY OF IMPORTANT TERMS (compiled by Dr Melanie Bussey)

Bernoulli's principle A low-pressure zone is created in a region of high fluid flow velocity, and a high-pressure zone is created in a region of low fluid flow velocity.

Boundary layer The thin layer of fluid that is adjacent to the surface of a body moving through the fluid. The fluid flow in the boundary layer may be **laminar flow** or **turbulent flow**.

Buoyancy force The upward force exerted on an immersed body by the fluid it displaces.

Centre of mass An imaginary balance point of a body; the point about which all of the mass particles of the body are evenly distributed. In the context of movement analysis, coincident with the centre of gravity.

Centre of percussion That point in a body moving about a fixed axis of rotation at which it may strike an obstacle without communicating an acceleration (or shock) to that axis.

Centre of pressure The effective point of application of a force distributed over a surface.

Collinear forces Forces whose lines of action are the same.

Damping Any effect that tends to reduce the amplitude of the oscillations of an oscillatory system.

Drag The mechanical force generated by a solid object travelling through a fluid that acts in the direction opposite to the movement of the object; any object moving through a fluid experiences drag. See also **lift**.

Ecological validity The methods, materials and setting of the experiment must approximate the real-life circumstances that are under study. See also **validity**.

Energy The capacity of a system to do work. See also **kinetic energy** and **potential energy**.

Equilibrium The state of a system whose acceleration is unchanged; a state of balance between various physical forces. See also **neutral equilibrium**, **stable equilibrium** and **unstable equilibrium**.

Force plate or platform A device that measures the contact force between an object and its surroundings, usually the ground reaction force.

Free body diagram A diagram in which the object of interest is isolated from its surroundings and all of the force vectors acting on the body are shown.

Integral The result of the process of **integration**; the area under a variable–time curve.

Integration The mathematical process of calculating an **integral**.

Kinetic energy The ability of a body to do work by virtue of its motion. See also **potential energy**.

Laminar flow At slow speeds the flow of a fluid smoothly over the surface of an object. The fluid flow can be considered as a series of thin plates (or laminae, hence the name) sliding smoothly past each other. 'Information' is passed between the elements of the fluid on a microscopic scale. See also **turbulent flow**.

Lift The component of fluid force that acts perpendicular to the direction of movement of an object through the fluid. Arises when the object deflects the fluid flow asymmetrically. See also **drag**.

Magnus effect The curve in the path of a spinning ball caused by a pressure differential around the ball.

Neutral equilibrium The state of a body in which the body will remain in a location if displaced from another location. See also **stable equilibrium** and **unstable equilibrium**.

Piezoelectric The ability of crystals to generate a voltage in response to an applied mechanical stress.

Potential energy The ability of a body to do work by virtue of its position. See also **kinetic energy**.

Shear force A force applied parallel to an object creating deformation internally in a direction at right angles to that of the force.

Signal amplitude A non-negative scalar measure of a wave's magnitude of oscillation, that is, the magnitude of the maximum disturbance in the medium during one wave cycle. See also **signal frequency**.

Signal frequency The measurement of the number of times a repeated event occurs per unit of time. It is also defined as the rate of change of phase of a sinusoidal waveform. See also **signal amplitude**.

Stable equilibrium The state of a body in which the body will return to its original location if it is displaced. See also **neutral equilibrium** and **unstable equilibrium**.

Steady-state response The response of a system at equilibrium. The steady-state response does not necessarily mean the response is a fixed value. See also **transient response**.

Stress Force per unit area.

Tangent (of an angle) The ratio of the length of the side opposite to an angle to the side adjacent to the angle in a right (or right-angle) triangle.

Tangential velocity The change in linear position along the instantaneous line tangent to the curve per unit of time of a body moving along a curved path.

Transient response The response of a system before achieving equilibrium. See also **steady-state response**.

Turbulent flow Fluid flow characterised by a series of vortices or eddies. 'Information' is passed between the elements of the fluid on a macroscopic scale. This is by far the predominant fluid flow in nature. See also **laminar flow**.

Unstable equilibrium The state of a body in which the displacement of the body continues to increase once it has been displaced. See also **neutral equilibrium** and **stable equilibrium**.

Validity Results that accurately reflect the concept being measured. An experiment is said to possess external validity if the experiment's results hold across different experimental settings, procedures and participants. An experiment is said to possess internal validity if it properly demonstrates a causal relation between two variables. See also **ecological validity**.

Viscosity The measure of a fluid's resistance to flow.

Wake The disruption of fluid flow downstream of a body caused by the passage of the body through the fluid.

Table 5.1 Calculation of the two-dimensional position of the whole body centre of mass: cadaver data adjusted to correct for fluid loss

SEGMENT	MASS FRACTION $m' = m/m$	LENGTH RATIO[1,2] lr	x-COORDINATE				y-COORDINATE			
			PROXIMAL x_p	DISTAL x_d	MASS CENTRE[3], x_m	MOMENT = $m'x_m$	PROXIMAL y_p	DISTAL y_d	MASS CENTRE[3], y_m	MOMENT = $m'y_m$
Head and neck	0.081	0.500								
Trunk	0.497	0.450								
Right upper arm	0.028	0.436								
Right forearm	0.016	0.430								
Right hand	0.006	0.506								
Right thigh	0.101	0.433								
Right calf	0.046	0.433								
Right foot	0.014	0.449								
Left upper arm	0.028	0.436								
Left forearm	0.016	0.430								
Left hand	0.006	0.506								
Left thigh	0.101	0.433								
Left calf	0.046	0.433								
Left foot	0.014	0.449								
Whole body	1.000	—	—	—	—	$= \Sigma m' x_m$	—	—	—	$= \Sigma m' y_m$

Notes:
1. Using simple ratios although many biomechanists use regression equations.
2. Defined as the distance of the centre of mass from the proximal point divided by the distance from the proximal point to the distal point. Joint centre locations are given in Box 6.2. The proximal point is the proximal joint and the distal point is the distal joint except for the following. For the head and neck, the 'proximal' point is the top (vertex) of the head and the distal point is the midpoint of a line joining the tip of the spinous process of the seventh cervical vertebra and the suprasternal notch (or of the line joining the two shoulder joint centres). For the trunk, the proximal point is the same as the distal point for the head and neck and the distal point is the midpoint of the line joining the hips. For the hand, the distal point is the third metacarpophalangeal joint (the third knuckle). For the foot, the proximal point is the most posterior point on the heel (calcaneus), and the distal point is the end of the second toe.
3. $x_m = x_p + lr\,(x_d - x_p)$; $y_m = y_p + lr\,(y_d - y_p)$.

FURTHER READING

Bartlett, R.M. (1999) *Sports Biomechanics: Reducing Injury and Improving Performance*, London: E & FN Spon. Chapter 4 provides a simple introduction to inverse dynamics without being too mathematical.

Lees, A. and Lake, M. (2007) Force and pressure measurement, in C.J. Payton and R.M. Bartlett (eds) *Biomechanical Evaluation of Movement in Sport and Exercise*, Abingdon: Routledge. An up-to-date coverage of force and pressure measurement in sport.

6 The anatomy of human movement

Knowledge assumed
Basic axes and planes of
movement (Chapter 1)
Analysis of sports movements
(Chapters 1 to 4)
Relationship between force
and movement (Chapter 5)
Basic muscle physiology (see
Further reading)

INTRODUCTION

In this chapter we will consider the anatomical principles that apply to movement in sport and exercise and how the movements of the sports performer are generated. Anatomy is an old branch of science, in which the use of Latin names is still routine in the English-speaking world. As most sports biomechanics students do not, understandably, speak Latin, the use of Latin words will be avoided, unless necessary, in this chapter; so, for example, Latin names of the various types of joint are not used. Where this avoidance is not possible, and this includes the naming of most muscles, some brief guidance to the grammar of this antique language is given. We shall also look at how electromyography can be used in the study of sports movements and the use of isokinetic dynamometry in recording muscle torques.

BOX 6.1 LEARNING OUTCOMES

After reading this chapter you should be able to:

- define the planes and axes of movement, and name and describe all of the principal movements in those planes in sport and exercise
- identify the functions of the skeleton and give examples of each type of bone
- describe typical surface features of bone and how these can be recognised superficially
- understand the tissue structures involved in the joints of the body and the factors contributing to joint stability and mobility
- identify the features of synovial joints and give examples of each class of these joints
- understand the features and structure of skeletal muscles
- classify muscles both structurally and functionally
- describe the types and mechanics of muscle contraction and appreciate how tension is produced in muscle
- understand how the total force exerted by a muscle can be resolved into components depending on the angle of pull
- appreciate the applications of electromyography to the study of sports skills
- understand why the recorded EMG differs from the physiological signal and how to use EMG measuring equipment in sports movements
- describe the main methods of quantifying the EMG signal in the time and frequency domains
- appreciate how and why isokinetic dynamometry is used to record the net muscle torque at a joint.

THE BODY'S MOVEMENTS

Movements of the human musculoskeletal system

As we noted in Chapter 1, a precise description of human movement requires the definition of a reference position or posture from which these movements are specified. The two positions used are the fundamental (Figure 1.2(a)) and anatomical (Figure 1.2(b)) reference positions. With the exception of the forearms and hands, the fundamental and anatomical reference positions are the same. The fundamental position of Figure 1.2(a) is similar to a 'stand to attention'. The forearm is in its neutral position, neither pronated ('turned in') nor supinated ('turned out'). In the anatomical position of Figure 1.2(b), the forearm has been rotated from the neutral position of Figure 1.2(a) so that the palm of the hand faces forwards. Movements of the hand and fingers are defined from this position, movements of the forearm (radioulnar joints) are defined from the fundamental reference position and movements at other joints can be defined from either.

Planes and axes of movement

As we noted in Chapter 1, movements at the joints of the human musculoskeletal system are mainly rotational and take place about a line perpendicular to the plane in which they occur. This line is known as an axis of rotation. Three axes – the sagittal, frontal and vertical (see also Box 1.2) – can be defined by the intersection of pairs of the planes of movement as in Figure 1.2. These movements are specified in detail and expanded upon in the next section.

Movements in the sagittal plane about the frontal axis

- Flexion, shown in Figure 1.3, is a movement away from the middle of the body in which the angle between the two body segments decreases – a 'bending' movement. The movement is usually to the anterior, except for the knee, ankle and toes. The term hyperflexion is sometimes used to describe flexion of the upper arm beyond the vertical. It is cumbersome and is completely unnecessary if the range of movement is quantified.
- Extension, also shown in Figure 1.3, is the return movement from flexion. Continuation of extension beyond the reference position is known, anatomically, as hyperextension. The return movement from a hyperextended position is usually called flexion in sports biomechanics although, in strict anatomical terms, it is described, somewhat cumbersomely, as reduction of hyperextension.
- Dorsiflexion and plantar flexion are normally used to define sagittal plane movements at the ankle joint. In dorsiflexion, the foot moves upwards towards the

anterior surface of the calf; in plantar flexion, the foot moves downwards towards the posterior surface of the calf.

Movements in the frontal plane about the sagittal axis

- Abduction, shown in Figure 1.4, is a sideways movement away from the middle of the body or, for the fingers, away from the middle finger.
- Radial flexion – also known as radial deviation – denotes the movement of the middle finger away from the middle of the body and can also be used for the other fingers.
- The term hyperabduction is sometimes used to describe abduction of the upper arm beyond the vertical.
- Adduction, also shown in Figure 1.4, is the return movement from abduction towards the middle of the body or, for the fingers, towards the middle finger.
- Ulnar flexion, also known as ulnar deviation, denotes the movement of the middle finger towards the middle of the body and can also be used for the other fingers.
- Continuation of adduction beyond the reference position is usually called hyperadduction. This is only possible when combined with some flexion.
- The return movement from a hyperadduction position is often called abduction in sports biomechanics although, in strict anatomical terms, it is described, again rather cumbersomely, as reduction of hyperadduction.
- Lateral flexion to the right or to the left, shown in Figure 6.1(a), is the sideways bending of the trunk to the right or left and, normally, the return movement from the opposite side.
- Eversion and inversion refer to the raising of the lateral and medial border of the foot (the sides of the foot furthest from and nearest to the middle of the body) with respect to the other border. Eversion cannot occur without the foot tending to be displaced into a toe-out, or abducted, position; likewise, inversion tends to be accompanied by adduction. The terms pronation and supination of the foot, shown in Figures 6.1(b) and (c), are widely used in describing and evaluating running gait, and may already be familiar to you for this reason. Pronation of the foot involves a combination of eversion and abduction (Figure 6.1(b)), along with dorsiflexion of the ankle. Supination involves inversion and adduction (Figure 6.1(c)) along with plantar flexion of the ankle. These terms should not be confused with pronation and supination of the forearm (see below). When the foot is bearing weight, as in running, its abduction and adduction movements are restricted by friction between the shoe and the ground. Medial and lateral rotation of the lower leg is then more pronounced than in the non-weight-bearing positions of Figures 6.1(b) and (c).

Figure 6.1 Movements in the frontal plane about the sagittal axis: (a) lateral flexion; (b) pronation; (c) supination.

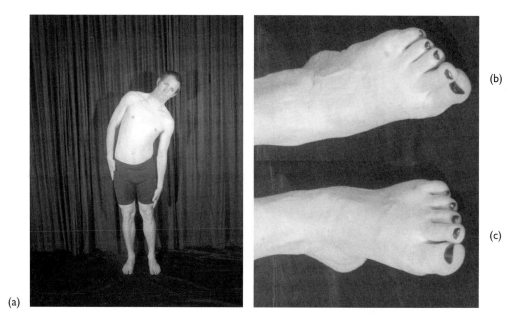

(b)

(c)

(a)

Movements in the horizontal plane about the vertical axis

- External and internal rotation, shown in Figure 1.5, are the outwards and inwards movements of the leg or arm about their longitudinal axes – these movements are also known, respectively, as lateral and medial rotation. External and internal rotation of the forearm are referred to, respectively, as supination and pronation.
- Rotation to the left and rotation to the right are the rather obvious terms for horizontal plane movements of the head, neck and trunk.
- Horizontal flexion and extension (or horizontal abduction and adduction), shown in Figure 1.6, define the rotation of the arm about the shoulder joint or the leg about the hip joint from a position of 90° abduction. In sports biomechanics, movements from any position in the horizontal plane towards the anterior are usually called horizontal flexion and those towards the posterior horizontal extension. (In strict anatomical terms, these movements are named from the 90° abducted position; the return movements towards that position are called reduction of horizontal flexion and extension respectively).

Movements of the thumb

The movements of the thumb, shown in Figures 6.2(a) to (c), may appear to be confusingly named.

227

Figure 6.2 Movements of the thumb: (a) abduction–adduction; (b) flexion–extension; (c) hyperflexion.

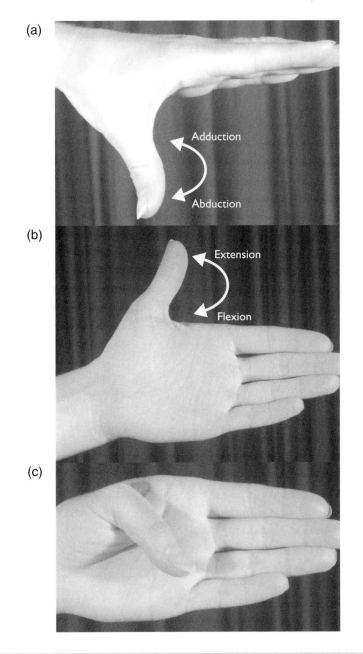

- Abduction and adduction (Figure 6.2(a)) are used to define movements away from and towards the palm of the hand in the sagittal plane; hyperadduction is the continuation of adduction beyond the starting position.

- Extension and flexion (Figure 6.2(b)) refer to frontal plane movements away from and towards the index finger; hyperflexion (Figure 6.2(c)) is the movement beyond the starting position.
- Opposition is the movement of the thumb to touch the tip of any of the four fingers of the same hand. It involves abduction and hyperflexion of the thumb.

Circumduction of the arm and leg

The movement of the arm or leg to describe a cone is called circumduction and is a combination of flexion and extension with abduction and adduction. Several attempts have been made to define movements in other diagonal planes but none has been adopted universally.

Movements of the shoulder girdle

The movements of the shoulder girdle are shown in Figure 6.3, along with the humerus movements with which they are usually associated.

- Elevation (shown in Figure 6.3(a)), and depression are upward and downward linear movements of the scapula. They are generally accompanied by some upward (Figure 6.3(b)) and downward scapular rotation, respectively, movements approximately in the frontal plane. These rotations are defined by the turning of the distal end of the scapula – that further from the middle of the body – with respect to the proximal end – that nearer the middle of the body.
- Protraction and retraction describe the movements of the scapula away from (Figure 6.3(c)) and towards the vertebral column. These are not simply movements in the frontal plane but also have anterior and posterior components owing to the curvature of the thorax.
- Posterior and anterior tilt are the upwards and downwards movement, respectively, of the inferior angle – the lower tip – of the scapula away from (Figure 6.3(d)) or towards the thorax.

Pelvic girdle movements

Changes in the position of the pelvis are brought about by the motions at the lumbosacral joint, between the lowest lumbar vertebra and the sacrum, and the hip joints. Movements at these joints, shown in Figure 6.4, permit the pelvis to tilt forwards, backwards and sideways (laterally) and to rotate horizontally.

- Forward tilt, from the position in Figure 6.4(a) to that in Figure 6.4(b), involves increased inclination in the sagittal plane about the frontal axis. This results from lumbosacral hyperextension and, in the standing position, hip flexion. The lower

Figure 6.3 Shoulder girdle movements: (a) elevation; (b) upward rotation; (c) abduction; (d) upward tilt.

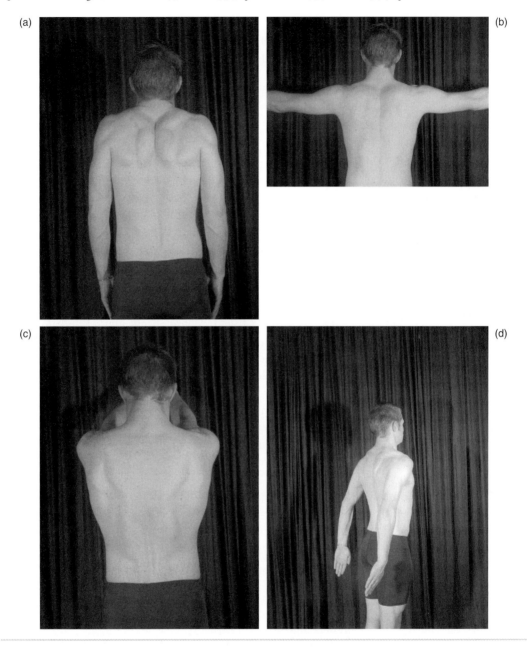

part of the pelvic girdle where the pubic bones join – the symphysis pubis – turns downwards and the posterior surface of the sacrum turns upwards.

- Backward tilt, from the position in Figure 6.4(a) to that in Figure 6.4(c), involves decreased inclination in the sagittal plane about the frontal axis. This results from

Figure 6.4 Pelvic girdle movements: (a) neutral sagittal plane position; (b) forward tilt; (c) backward tilt; (d) rotation to the left.

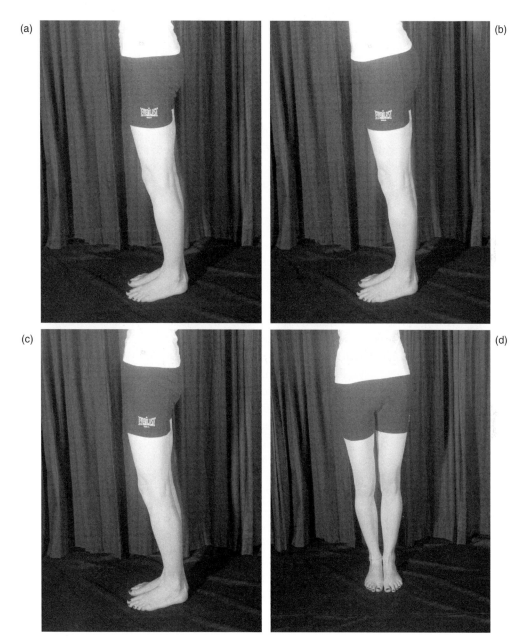

lumbosacral flexion and, in the standing position, hip extension. The symphysis pubis moves forwards and upwards and the posterior surface of the sacrum turns somewhat downwards.

- Lateral tilt is the movement of the pelvis in the frontal plane about the sagittal axis such that one iliac crest is lowered and the other is raised. This can be demonstrated by standing on one foot with the other slightly raised directly upwards off the ground, keeping the leg straight. The tilt is named from the side of the pelvis that moves downwards; in lateral tilt of the pelvis to the left, the left iliac crest is lowered and the right is raised. This is a combination of right lateral flexion of the lumbosacral joint, abduction of the left hip and adduction of the right.
- Rotation or lateral twist (Figure 6.4(d)) is the rotation of the pelvis in the horizontal plane about a vertical axis. The movement is named after the direction towards which the front of the pelvis turns.

THE SKELETON AND ITS BONES

In this section, the functions of the human skeleton and the form, nature and composition of its bones will be considered. There are 206 bones in the human skeleton, of which 177 engage in voluntary movement. The functions of the skeleton are: to protect vital organs such as the brain, heart and lungs; to provide rigidity for the body; to provide muscle attachments whereby the bones function as levers, allowing the muscles to move them about the joints; to enable the manufacture of blood cells; and to provide a storehouse for mineral metabolism.

The skeleton, shown in Figure 6.5, is often divided into the axial skeleton, comprising the skull, lower jaw, vertebrae, ribs, sternum, sacrum and coccyx, which is mainly protective, and the appendicular skeleton, consisting of the shoulder girdle and upper extremities, and the pelvic girdle and lower extremities, which functions in movement.

Bone structure

Macroscopically, there are two types of bone tissue: cortical, or compact, bone and trabecular, or cancellous, bone. The first forms the outer shell, or cortex, of a bone and has a dense structure. The second has a loose latticework structure of trabeculae or cancelli; the spaces – or interstices – between the trabeculae are filled with red marrow in which red blood cells form. Cancellous bone tissue is arranged in concentric layers, or lamellae, and its cells, called osteocytes, are supplied with nutrients from blood vessels passing through the red marrow. The lamellar pattern and material composition of the two bone types are similar. Different porosity is the principal distinguishing feature, and the distinction between the two types might be considered somewhat arbitrary. Biomechanically, the two types of bone should be considered as one material

Figure 6.5 The skeleton.

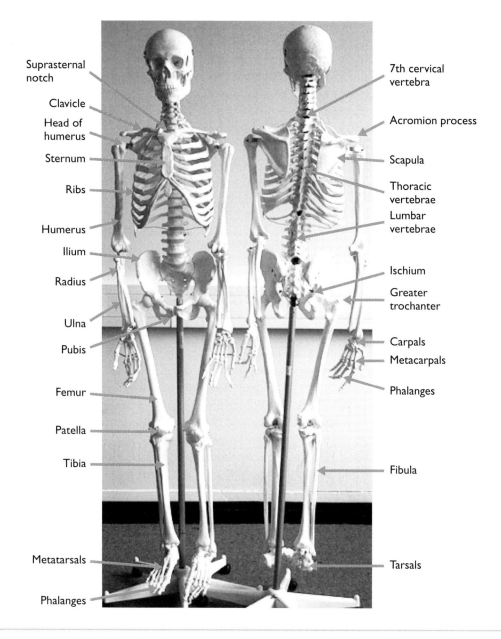

Suprasternal notch

Clavicle

Head of humerus

Sternum

Ribs

Humerus

Ilium

Radius

Ulna

Pubis

Femur

Patella

Tibia

Metatarsals

Phalanges

7th cervical vertebra

Acromion process

Scapula

Thoracic vertebrae

Lumbar vertebrae

Ischium

Greater trochanter

Carpals

Metacarpals

Phalanges

Fibula

Tarsals

with a wide range of densities and porosities. It can be classified as a composite material in which the strong but brittle mineral element is embedded in a weaker but more ductile one consisting of collagen and ground substance. Like many similar but inorganic composites, such as carbon-reinforced fibres, which are important in sports equipment, this structure gives a material whose strength to weight ratio exceeds that of either of its constituents.

With the exception of the articular surfaces, bones are wholly covered with a membrane known as the periosteum. This has a strong outer layer of fibres of a protein, collagen, and a deep layer that produces cells called osteoblasts, which participate in the growth and repair of the bone. The periosteum also contains capillaries, which nourish the bone, and it has a nerve supply. It is sensitive to injury and is the source of much of the pain from fracture, bone bruises and shin splints. Muscles generally attach to the periosteum, not directly to the bone, and the periosteum attaches to the bone by a series of root-like processes.

Classification of bones

Bones can be classified according to their geometrical characteristics and functional requirements, as follows.

- Long bones occur mostly in the appendicular skeleton and function for weight-bearing and movement. They consist of a long, central shaft, known as the body or diaphysis, the central cavity of which consists of the medullary canal, which is filled with fatty yellow marrow. At the expanded ends of the bone, the compact shell is very thin and the trabeculae are arranged along the lines of force transmission. In the same region, the periosteum is replaced by smooth, hyaline articular cartilage. This has no blood supply and is the residue of the cartilage from which the bone formed. Examples of long bones are the humerus, radius and ulna of the upper limb, the femur, tibia and fibula of the lower limb, and the phalanges (Figure 6.5).
- Short bones are composed of cancellous tissue and are irregular in shape, small, chunky and roughly cubical; examples are the carpal bones of the hand and the tarsal bones of the foot.
- Flat bones are basically a sandwich of richly veined cancellous bone within two layers of compact bone. They serve as extensive flat areas for muscle attachment and, except for the scapulae, enclose body cavities. Examples are the sternum, ribs and ilium.
- Irregular bones are adapted to special purposes and include the vertebrae, sacrum, coccyx, ischium and pubis.
- Sesamoid bones form in certain tendons; the most important example is the patella (the kneecap).

The surface of a bone

The surface of a bone is rich in markings that show its history; some examples are shown in Figure 6.6. Some of these markings can be seen at the skin surface and many others can be easily felt (palpated) – some of these are the anatomical landmarks used to estimate the axes of rotation of the body's joints (see Box 6.2), which are crucially important for quantitative biomechanical investigations of human movements in sport and exercise.

- Lumps on bones known as tuberosities or tubercles – the latter are smaller than the former – and projections, known as processes, show attachments of strong fibrous cords, such as tendons. The styloid processes (Figure 6.6(a)) on the radius and ulna are used to locate the wrist axis of rotation in the sagittal plane. The acromion process on the scapula (Figure 6.5) is often used to estimate the shoulder joint axis of rotation in the sagittal plane (see Box 6.2).
- Ridges and lines indicate attachments of broad sheets of fibrous tissue known as aponeuroses or intermuscular septa.
- Grooves, known as furrows or sulci, holes – or foramina, notches and depressions or fossae – often suggest important structures, for example grooves for tendons. The suprasternal notch (Figure 6.5) is often used to establish the trunk–neck boundary for centre of mass calculations (see Table 5.1).
- A projection from a bone was defined above as a process, while a rounded prominence at the end of a bone is termed a condyle, the projecting part of which is sometimes known as an epicondyle. The humeral epicondyles (Figure 6.6(b)) and the femoral condyles (Figure 6.6(c)) are used, respectively, to locate the elbow and knee axes of rotation in the sagittal plane.
- Special names are given to other bony prominences. The greater trochanter on the lateral aspect of the femur, shown in Figure 6.5, is often used to estimate the position of the hip joint centre. The medial and lateral malleoli, on the medial and lateral aspects of the distal end of the tibia (Figure 6.6(d)), are used to locate the ankle axis of rotation in the sagittal plane.

Bone fracture

The factors affecting bone fracture, such as the mechanical properties of bone and the forces to which bones in the skeleton are subjected during sport or exercise, will not be discussed in detail here. The loading of living bone is complex because of the combined nature of the forces, or loads, applied and because of the irregular geometric structure of bones. For example, during the activities of walking and jogging, the tensile and compressive stresses along the tibia are combined with transverse shear stresses, caused by torsional loading associated with lateral and medial rotation of the tibia. Although the tensile stresses are, as would be expected, much larger for jogging than walking, the shear stresses are greater for the latter activity. Most fractures are produced by such a combination of several loading modes.

After fracture, bone repair is effected by two types of cell, known as osteoblasts and osteoclasts. Osteoblasts deposit strands of fibrous tissue, between which ground substance is later laid down, and osteoclasts dissolve or break up damaged or dead bone.

Figure 6.6 Surface features of bones: (a) at the wrist; (b) at the elbow; (c) at the knee; (d) at the ankle.

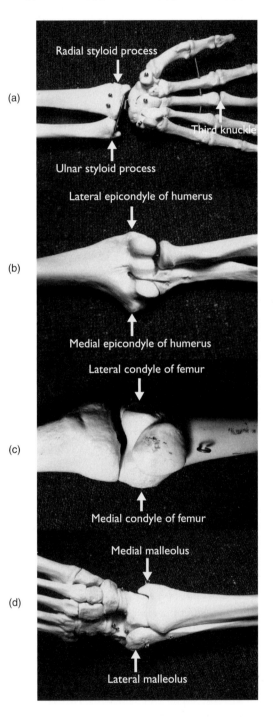

BOX 6.2 LOCATION OF MAIN JOINT SAGITTAL AXES OF ROTATION AND JOINT CENTRES OF ROTATION (see also Table 5.1)

Joint	Sagittal axis	Joint centre
Shoulder	About 10% of the distance between the most lateral point on the lateral border of the acromion process and the elbow joint axis, below the acromion process (Figure 6.5).	At the centre of the head of the humerus (cannot be palpated).
Elbow	Along line joining the medial and lateral epicondyles of the humerus (Figure 6.6(b)).	Midway along that line.
Wrist	Along line joining styloid processes on radius and ulna (Figure 6.6(a)).	Midway along that line.
Hip	Approximately along line joining the most proximal palpable point on the greater trochanter on the femur to the equivalent point on the contralateral femur (perhaps 1% of the distance from this point to the knee axis of rotation, superior to the greater trochanter). See Figure 6.5.	At the centre of the head of the femur (cannot be palpated).
Knee	Along line joining medial and lateral condyles on femur (Figure 6.6(c)).	Midway along that line.
Ankle	Along line joining most distal palpable point of tibial (medial) malleolus and most lateral point on fibular (lateral) malleolus.	Midway along that line.

Initially, when the broken ends of the bone are brought into contact, they are surrounded by a mass of blood. This is partly absorbed and partly converted, first into fibrous tissue then into bone. The mass around the fractured ends is called the callus. This forms a thickening, for a period of months, which will gradually be smoothed away, unless the ends of the bone have not been correctly aligned. In that case, the callus will persist and form an area where large mechanical stresses may occur, rendering the region susceptible to further fracture.

THE JOINTS OF THE BODY

Joints, also known as articulations, occur between the bones or cartilage of the skeleton. They allow free movement of the various parts of the body or more restricted movements, for example during growth or childbirth. Other tissues that may be present in the joints of the body are dense, fibrous connective tissue, which includes ligaments, and synovial membrane. The nature and biomechanical functions of these and other structures associated with joints will not be considered here.

Joint classification

Overall, joints are classified according to the movement they allow. In fibrous joints, the edges of the bone are joined by thin layers of the fibrous periosteum, as in the suture joints of the skull, where movement is undesirable. With age these joints disappear as the bones fuse. These are immovable joints. Ligamentous joints occur between two bones. The bones can be close together, as in the interosseus talofibular ligament, which allows only a little movement, or further apart, as in the broad and flexible interosseus membrane of fibrous tissue between the ulna and radius, which permits free movement. These are not true joints and are classed as slightly movable. Cartilaginous joints either consist of hyaline cartilage, as in the joint between the sternum and first rib, or fibro-cartilage, as in the intervertebral discs. Cartilaginous joints are classed as slightly movable.

The third classification of joint has a joint cavity surrounded by a sleeve of fibrous tissue, the ligamentous joint capsule, which unites the bones. Friction between the bones is minimised by smooth hyaline cartilage. Although cartilage has been traditionally regarded as a shock absorber, this role is now considered unlikely. The functions of the cartilage in such joints are mainly to help to reduce stresses between the contacting surfaces, by widely distributing the joint loads, and to allow movement with minimal friction. The inner surface of the capsule is lined with the delicate synovial membrane, the cells of which exude the synovial fluid that lubricates the joint. This fluid converts potentially compressive solid stresses into equally distributed hydrostatic ones, and nourishes the bloodless hyaline cartilage. These freely moveable or synovial joints (Figure 6.7) are the most important in human movement. The changing relationship of the bones to each other during movement creates spaces filled by synovial folds and fringes attached to the synovial membrane. When filled with fat cells these are called fat pads. In certain synovial joints, such as the sternoclavicular and distal radioulnar joints, fibrocartilaginous discs occur that wholly or partially divide the joint. Synovial joints can be classified as follows, based mainly on how many degrees of rotational freedom the joint allows (Figure 6.7).

- Plane joints (also known as gliding or irregular joints) are joints in which only slight gliding movements occur. These joints have an irregular shape. Examples are the intercarpal joints of the hand (Figure 6.7(a)), the intertarsal joints of the foot, the acromioclavicular joint and the heads and tubercles of the ribs. The joints are classed in the literature both as non-axial, because they glide more or less on a plane surface, and multiaxial. The latter term is presumably used because the surfaces are not plane but have a large radius of curvature and, therefore, an effective centre of rotation

Figure 6.7 Classification of synovial joints: (a) plane joint; (b) hinge joint; (c) pivot joint; (d) condyloid joint; (e) saddle joint; (f) ball and socket joint.

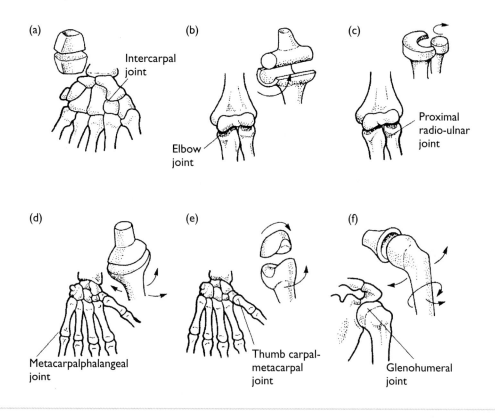

some considerable distance from the bone. The former description seems more useful. Although individual joint movements are small, combinations of several such joints, as in the carpal region of the hand, can result in significant motion.

- Hinge joints are joints in which the concave surface of one bone glides partially around the convex surface of the other. Examples are the elbow (Figure 6.7(b)) and ankle joints and the interphalangeal joints of the fingers and toes. The knee is not a simple hinge joint, although it appears that way when bearing weight. Hinge joints are uniaxial, permitting only the movements of flexion and extension.

- Pivot joints are joints in which one bone rotates about another. This may involve the bones fitting together at one end, with one rotating about a peg-like pivot in the other, as in the atlanto-axial joint between the first and second cervical vertebrae. The class is also used to cover two long bones lying side by side, as in the proximal radioulnar joint of Figure 6.7(c). These joints are uniaxial, permitting rotation about the vertical axis in the horizontal plane.

- Condyloid joints are classified as biaxial joints, permitting flexion–extension and abduction–adduction (and, therefore, circumduction). The class is normally used to cover two slightly different types of joint. One of these has a spheroidal surface that articulates with a spheroidal depression, as in the metacarpophalangeal or 'knuckle' joints (Figure 6.7(d)) – 'condyloid' means 'knuckle-like'. These joints are potentially triaxial but lack the musculature to perform rotation about a vertical axis. The other type, which is sometimes classified separately as ellipsoidal joints, is similar in most respects except that the articulating surfaces are ellipsoidal rather than spheroidal, as in the wrist joint.

- Saddle joints consist of two articulating saddle-shaped surfaces, as in the thumb carpometacarpal joint, shown in Figure 6.7(e). These are biaxial joints, with the same movements as other biaxial joints but with greater range.
- Ball and socket joints, also known as spheroidal joints, have the spheroidal head of one bone fitting into the cup-like cavity of the other, as in the hip joint and the shoulder (or glenohumeral) joint. The latter is shown in Figure 6.7(f). These are triaxial joints, permitting movements in all three planes.

Joint stability and mobility

The stability, or immobility, of a joint is the joint's resistance to displacement. It depends on the following factors:

- The shape of the bony structure, including the type of joint and the shape of the bones. This is a major stability factor in some joints, such as the elbow and hip, but of far less importance in others, for example the knee and shoulder joints.
- The ligamentous arrangement, including the joint capsule, which is crucial in, for example, the knee joint.
- The arrangement of fascia, tendons and aponeuroses.
- Position – joints are more stable in the close-packed position, with maximal contact between the articular surfaces and with the ligaments taut, than in a loose-packed position.
- Atmospheric pressure, providing it exceeds the pressure within the joint, as in the hip joint.
- Muscular contraction – depending on the relative positions of the bones at a joint, muscles may have a force component capable of pulling the bone into the joint (see Figure 6.17); this is particularly important when the bony structure is not inherently stable, as in the shoulder joint.

Joint mobility or flexibility is widely held to be desirable for sportsmen and sports-women. It is usually claimed to reduce injury. Although this is probably true, excessive mobility can sacrifice important stability and predispose to injury. It is also sometimes claimed that improved mobility enhances performance; although this is impeccably logical it is not well substantiated. Mobility is highly joint-specific and is affected by body build, heredity, age, sex, fitness and exercise. Participants in sport and exercise are usually more flexible than non-participants owing to the use of joints through greater ranges, avoiding adaptive shortening of muscles. Widely differing values are reported in the scientific literature for normal joint ranges of movement. The discrepancies may be attributable to unreliable instrumentation, lack of standard experimental protocols and inter-individual differences.

MUSCLES – THE POWERHOUSE OF MOVEMENT

Muscles are structures that convert chemical energy into mechanical work and heat energy. In studying sport and exercise movements biomechanically, the muscles of interest are the skeletal muscles, used for moving and for posture. This type of muscle has striated muscle fibres of alternating light and dark bands. Muscles are extensible, that is they can stretch or extend, and elastic, such that they can resume their resting length after extending. They possess excitability and contractility. Excitability means that they respond to a chemical stimulus by generating an electrical signal, the action potential, along the plasma membrane. Contractility refers to the unique ability of muscle to shorten and hence produce movement. Skeletal muscles account for approximately 40–50% of the mass of an adult of normal weight. From a sport or exercise point of view, skeletal muscles exist as about 75 pairs. The main skeletal muscles are shown in Figure 6.8. The proximal attachment of a muscle, that nearer the middle of the body, is known as the origin and the distal attachment as the insertion. The attachment points of skeletal muscles to bone and the movements they cause are not listed here but can be found in many books dealing with exercise physiology and anatomy (for example, Marieb, 2003; see Further Reading, page 280) as well as on this book's website.

Muscle structure

Each muscle fibre is a highly specialised, complex, cylindrical cell. The cell is elongated and multinucleated, 0.01–0.1 mm in diameter and seldom more than a few centimetres long. The cytoplasm of the cell is known as the sarcoplasm. This contains large amounts of stored glycogen and a protein, myoglobin, which is capable of binding oxygen and is unique to muscle cells. Each fibre contains many smaller, parallel elements, called myofibrils, which run the length of the cell and are the contractile components of skeletal muscle cells. The sarcoplasm is surrounded by a delicate plasma membrane called the sarcolemma. The sarcolemma is attached at its rounded ends to the endomysium, the fibrous tissue surrounding each fibre. Units of 100–150 muscle fibres are bound in a coarse, collagenic fibrous tissue, the perimysium, to form a fascicle. The fascicles can be much longer than individual muscle fibres, for example around 250 mm long for the hamstring muscles. Several fascicles are bound into larger units enclosed in a covering of yet coarser, dense fibrous tissue, the deep fascia, or epimysium, to

Figure 6.8 Main skeletal muscles: (a) anterior view; (b) posterior view (adapted from Marieb, 2003; see Further Reading, page 280).

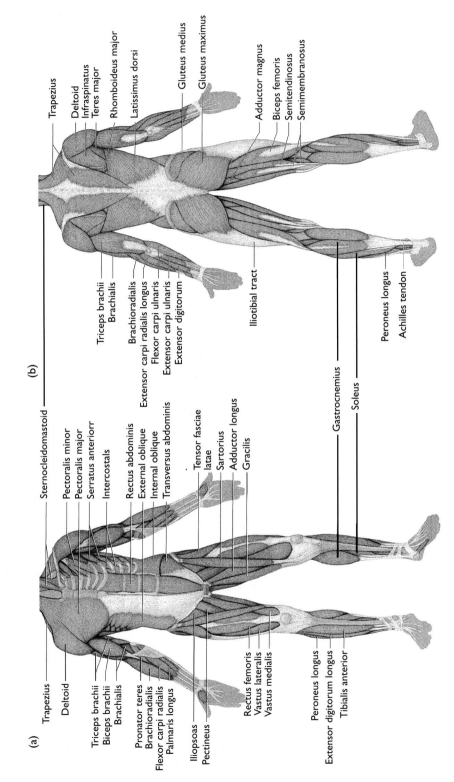

form muscle. The epimysium separates the muscle from its neighbours and facilitates frictionless movement.

The contractile cells are concentrated in the soft, fleshy central part of the muscle, called the belly. Towards the ends of the muscle the contractile cells finish but the perimysium and epimysium continue to the bony attachment as a cord-like tendon or flat aponeurosis, in which the fibres are plaited to distribute the muscle force equally over the attachment area. If the belly continues almost to the bone, then individual sheaths of connective tissue attach to the bone over a larger area.

The strength and thickness of the muscle sheath varies with location. Superficial muscles, particularly near the distal end of a limb have a thick sheath with additional protective fascia. The sheaths form a tough structural framework for the semi-fluid muscle tissue. They return to their original length even if stretched by 40% of their resting length. Groups of muscles are compartmentalised from others by intermuscular septa, usually attached to the bone and to the deep fascia that surrounds the muscles.

Muscle activation

Each muscle fibre is innervated by cranial or spinal nerves and is under voluntary control. The terminal branch of the nerve fibre ends at the neuromuscular junction or motor end-plate, which touches the muscle fibre and transmits the nerve impulse to the sarcoplasm. Each muscle is entered from the central nervous system by nerves that contain both motor and sensory fibres, the former of which are known as motor neurons. As each motor neuron enters the muscle, it branches into terminals, each of which forms a motor end-plate with a single muscle fibre. The term motor unit is used to refer to a motor neuron and all the muscle fibres that it innervates, and these can be spread over a wide area of the muscle. The motor unit can be considered the fundamental functional unit of neuromuscular control. Each nerve impulse causes all the muscle fibres of the motor unit to contract fully and almost simultaneously. The number of fibres per motor unit is sometimes called the innervation ratio. This ratio can be less than 10 for muscles requiring very fine control and greater than 1000 for the weight-bearing muscles of the lower extremity. Muscle activation is regulated through motor unit recruitment and the motor unit stimulation rate (or rate-coding). The former is an orderly sequence based on the size of the motor neuron. The smaller motor neurons are recruited first; these are typically slow twitch with a low maximum tension and a long contraction time. If more motor units can be recruited, then this mechanism dominates; smaller muscles have fewer motor units and depend more on increasing their stimulation rate.

Naming muscles

As noted in the introduction to this chapter, in the scientific literature muscles are nearly always known by their Latin names. The full name is musculus, which is often omitted or abbreviated to m or M., followed by adjectives or genitives of nouns. The name may refer to role, location, size or shape of the muscle. An example of a muscle named after its location is the latissimus dorsi – the broadest (latissimus) muscle of the back. The flexor digitorum profundus is named after its role – the deep (profundus) flexor of the fingers. The trapezius muscle – the English name is identical to the Latin one for this, but for only a few other, muscles – is named after its trapezoidal shape. Some English names have become accepted, such as the anterior deltoid; obvious English translations of the Latin names, for example the deep flexor of the fingers (see above) are – somewhat sadly in my view – not commonly encountered in most scientific literature.

Muscles are often described by their role, such as the flexors of the knee and the abductors of the humerus. Most muscles have more than one role in movement; multi-joint muscles have roles at more than one joint.

Structural classification of muscles

The internal structure or arrangement of the muscle fascicles is related to both the force of contraction and the range of movement and, therefore, serves as a logical way of classifying muscles. There are two basic types each of which is further subdivided.

Collinear muscles (Figures 6.9(a) to (e)) have muscle fascicles that are more or less parallel. A collinear muscle is capable of shortening by about one-third to one-half of its belly's length. Such muscles have a large range of movement, which is limited by the fraction of the muscle length that is tendinous. These muscles are very common in the extremities and are further divided as follows.

- Longitudinal muscles consist of long, strap-like fascicles parallel to the long axis, as shown schematically in Figure 6.9(a); examples are the rectus abdominis muscle of the abdominal wall and the sartorius, the longest muscle in the human body, which crosses the hip and knee joints across the front of the thigh.
- Quadrate muscles are four-sided, usually flat, with parallel fascicles, as in Figures 6.9(b) and (c). They may have a rhomboid shape as in the schematic representation of the rhomboideus major – a muscle of the scapula – in Figure 6.9(b), or rectangular, as, for example, the pronator quadratus, located on the anterior aspect of the forearm near the wrist and shown schematically in Figure 6.9(c).
- Fan-shaped muscles (Figure 6.9(d)) are relatively flat with almost parallel fascicles that converge towards the insertion point. A good example is the pectoralis major muscle on the upper anterior surface of the trunk.
- Fusiform muscles are usually rounded, tapering at either end (Figure 6.9(e)), and include the elbow flexors: brachialis, brachioradialis and biceps brachii. The location of the last of these is certainly familiar to most, if not all, sport and exercise science students. The brachialis lies directly underneath the biceps brachii, is a single joint muscle and is the main elbow flexor.

Figure 6.9 Structural classification of muscles. Collinear muscles: (a) longitudinal; (b) quadrate rhomboidal; (c) quadrate rectangular; (d) fan-shaped; (e) fusiform. Pennate muscles: (f) unipennate; (g) bipennate; (h) multipennate.

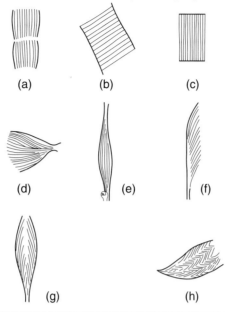

Pennate, or penniform, muscles (Figures 6.9(f) to (h)), have shorter fascicles than collinear muscles; the fascicles are angled away from an elongated tendon. This arrangement allows more fibres to be recruited, which provides a stronger, more powerful muscle at the expense of range of movement and speed of the limb moved. They account for 75% of the body's muscles, mostly in the large muscle groups, including the powerful muscles of the lower extremity. This classification is further divided into the following groups:

- Unipennate muscles lie to one side of the tendon, extending diagonally as a series of short, parallel fascicles, as in Figure 6.9(f); the tibialis posterior muscle of the ankle is an example.
- Bipennate muscles (Figure 6.9(g)) have a long central tendon, with fascicles in diagonal pairs on either side. This group includes the rectus femoris muscle of the thigh and the flexor hallucis longus, which flexes the big toe (hallux).
- Multipennate muscles converge to several tendons, giving a herringbone effect (Figure 6.9(h)), for example the deltoid (deltoideus in Latin).

Types of muscle contraction

The term 'muscle contraction' refers to the development of tension within the muscle. The term is a little confusing, as contraction means becoming smaller in much English usage. Some sport and exercise scientists would prefer the term 'action' to be used instead, but as this has yet to be widely adopted, I will use muscle contraction, of which there are three main types:

- In isometric, or static, contraction, the muscle develops tension with no change in overall muscle length, as when holding a dumbbell stationary in a biceps curl.
- In concentric contraction, the muscle shortens as tension is developed, as when a dumbbell is raised in a biceps curl.
- In eccentric contraction, the muscle develops tension while it lengthens, as in the lowering movement in a biceps curl.

Both concentric and eccentric contractions can, theoretically, be at constant tension (isotonic) or constant speed (isokinetic). However, most contractions normally involve neither constant tension nor constant speed.

Group action of muscles

When a fibre or muscle develops tension both ends tend to move; whether these movements actually occur depends on the resistance to movement and on the activity of other muscles. Furthermore, when a muscle develops tension, it tends to perform all of its possible actions at all joints it crosses. Because of these axioms, muscles act together rather than individually to bring about the movements of the human body, with each muscle playing a specific role – this is one important feature of coordinated movement. In such group actions, muscles are classified according to their role, as follows. Please note that the muscle group terminology used here is that normally used in sports biomechanics; however, many anatomists now prefer different names for these muscle group roles.

- The muscles that directly bring about a movement by contracting concentrically are known as the agonists, which means 'movers'. This group is sometimes divided into prime movers, which always contract to cause the movement, and assistant movers, which only contract against resistance or at high speed. However, electromyography (see below) does not usually support such a simple distinction. If we accept such a distinction, then brachialis and biceps brachii would be prime movers for elbow flexion while brachioradialis would be generally considered to be an assistant mover. This distinction is closely related to the idea of spurt and shunt muscles touched on later in this section.
- Antagonists are muscles that cause the opposite movement from that of specified agonists. Their normal role in group action is to relax when the agonists contract,

although there are many exceptions to this. At the elbow, the triceps brachii is antagonistic to brachialis and biceps brachii.

- Stabilisers contract statically to fix one bone against the pull of the agonists so that the bone at the other end can move effectively. Muscles that contract statically to prevent movements caused by gravity are sometimes called supporting muscles, such as the abdominal muscles in push-ups.
- Neutralisers prevent undesired actions of the agonists when the agonists have more than one function. They may do this by acting in pairs, as mutual neutralisers, when they enhance the required action and cancel the undesired ones. For example, the flexor carpi radialis flexes and abducts the wrist while the flexor carpi ulnaris flexes and adducts the wrist: acting together they produce only flexion. Such muscles are also sometimes called helping synergists. Neutralisers may also contract statically to prevent an undesired action of agonists that cross more than one joint (multi-joint muscles). The flexion of the fingers while the wrist remains in its neutral anatomical position involves static contraction of the wrist extensors to prevent the finger flexors from flexing the wrist. Such muscles are also sometimes called true synergists.

BOX 6.3 A SCHEMATIC MODEL OF SKELETAL MUSCLE

A simple schematic model of skeletal muscle is often used to represent its functionally different parts. The model used is normally similar to Figure 6.10, and has contractile, series elastic, and parallel elastic elements. The contractile component is made up of the myofibril protein filaments of actin and myosin and their associated coupling mechanism.

Figure 6.10 Simple schematic model of skeletal muscle.

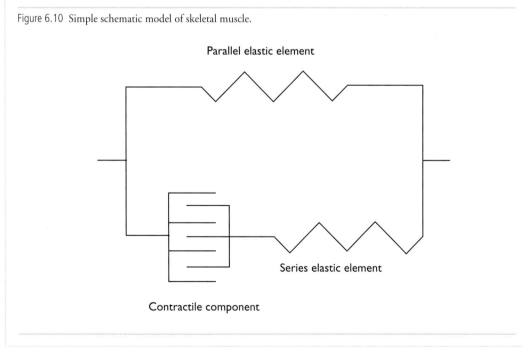

Parallel elastic element

Series elastic element

Contractile component

The series elastic element lies in series with the contractile component and transmits the tension produced by the contractile component to the attachment points of the muscle. The tendons account for by far the major part of this series elasticity, with elastic structures within the muscle cells contributing the remainder. The parallel elastic element comprises the epimysium, perimysium, endomysium and sarcolemma. The elastic elements store elastic energy when they are stretched and release this energy when the muscle recoils. The series elastic element is more important than the parallel elastic element in this respect. The elastic elements are important as they keep the muscle ready for contraction and ensure the smooth production and transmission of tension during contraction. They also ensure the return of the contractile component to its resting position after contraction. They may also help to prevent the passive overstretching of the contractile component when relaxed, reducing the risk of injury. In practice, the series and parallel elastic elements are viscoelastic rather than simply elastic, enabling them to absorb energy at a rate proportional to that at which force is applied and to dissipate energy at a rate that is time-dependent; this would require the addition of a damping element to each elastic element in Figure 6.10. In toe-touching, the initial stretch is elastic followed by a further elongation of the muscle–tendon unit owing to its viscosity.

The mechanics of muscular contraction

This section will consider the gross mechanical response of a muscle to various neural stimuli. Much of this information is derived from *in vitro*, electrical stimulation of the frog gastrocnemius. However, we will assume that similar responses occur *in vivo* for the stimulation of human muscle by motor nerves. Although each muscle fibre can only respond in an all-or-none way, a muscle contains many fibres and can contract with various force and time characteristics.

The muscle twitch

The muscle twitch is the mechanical response of a muscle to a single, brief, low-intensity stimulus. The muscle contracts and then relaxes, as represented in Figure 6.11(a). After stimulation, there is a short period of a few milliseconds when excitation–contraction coupling occurs and no tension is developed. This can be considered as the time to take up the slack in the elastic elements and is known as the latency (or latent) period. The contraction time is the time from onset of tension development to peak tension (Figure 6.11(a)) and lasts from 10 to 100 ms, depending on the make-up of the muscle fibres. If the tension developed exceeds the resisting load, the muscle will shorten. During the following relaxation time the tension drops to zero. If the muscle had shortened, it now returns to its initial length. The muscle twitch is normally a laboratory rather than an *in vivo* event. In most human movement, contractions are long and smooth and variations of the response are referred to as graded responses. These are regulated by two neural control mechanisms. The first – increasing the stimulation rate – involves increasing the rapidity of stimulation to produce wave summation. The second – increasing motor unit recruitment – involves recruitment of increasingly more motor units to produce multiple motor unit summation.

Wave summation and tetanus

The duration of an action potential is only a few milliseconds, which is very short compared with the following twitch. It is therefore possible for a series of action potentials, known as an action potential train, to be initiated before the muscle has completely relaxed. As the muscle is still partially contracted, the tension developed as a result of the second stimulus produces greater shortening than the first, as seen in Figures 6.11(b) and (c). The contractions are additive and the phenomenon is called wave summation. Increasing the stimulation rate will result in greater tension development as the relaxation time decreases until it eventually disappears. When this occurs, a smooth, sustained contraction results (Figure 6.11(d)) called tetanus; this is the normal form of muscle contraction in the body. It should be noted that prolonged tetanus leads to an eventual inability to maintain the contraction and a decline in the tension to zero. This condition is termed muscle fatigue.

Multiple motor unit summation

The wide gradation of contractions within muscles is achieved mainly by the differing activities in their various motor units – in stimulation rate and in the number of units recruited. The repeated, asynchronous twitching of all the recruited motor units leads

Figure 6.11 Muscle responses: (a) muscle twitch; (b) wave summation; (c) incomplete and (d) complete tetanus (adapted from Marieb, 2003; see Further Reading, page 280).

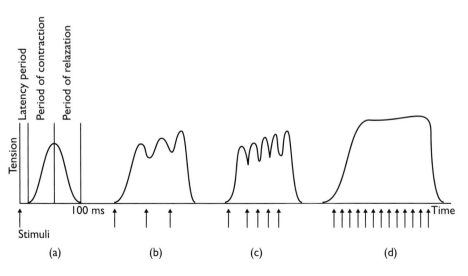

to brief summations or longer subtetanic or tetanic contractions of the whole muscle. For precise but weak movements only a few motor units will be recruited while far more will be recruited for forceful contractions. The smallest motor units with the fewest muscle fibres, normally type I (see below), are recruited first. The larger motor units, normally type IIA, then type IIB (see below) are activated only if needed. Both wave summation and multiple motor unit summation are factors in producing the smooth movements of skeletal muscle. Multiple motor unit summation is primarily responsible for the control of the force of contraction.

Treppe

The initial contractions in the muscle are relatively weak, only about half as strong as those that occur later. The tension development then has a staircase pattern called treppe, which is related to the suddenly increased availability of calcium ions. This effect, along with the increased enzymatic activity, the increase in conduction velocity across the sarcolemma, and the increased elasticity of the elastic elements, which are all consequent on the rise in muscle temperature, leads to the pattern of increasingly strong contractions with successive stimuli. The effect could be postulated as a reason for warming up before an event, but this view is not universally accepted.

Development of tension in a muscle

The tension developed in a muscle depends upon:

* The number of fibres recruited and their firing (or stimulation) rate and synchrony.
* The relative size of the muscle – the tension is proportional to the physiological cross-sectional area of the muscle; about 0.3 N force can be exerted per square millimetre of cross-sectional area.
* The temperature of the muscle and muscle fatigue.
* The pre-stretch of the muscle – a muscle that develops tension after being stretched (the stretch–shortening cycle, see below) performs more work because of elastic energy storage and other mechanisms; the energy is stored mostly in the series elastic elements but also in the parallel elastic ones.
* The mechanical properties of the muscle, as expressed by the length–tension, force–velocity and tension–time relationships (see below).

It should be noted that there are distinct differences in the rates of contraction, tension development and susceptibility to fatigue of individual muscle fibres. The main factor here is the muscle fibre type. Slow-twitch, oxidative type I fibres are suited for prolonged, low-intensity effort as they are fatigue-resistant because of their aerobic metabolism. However, they produce little tension as they are small and contract only slowly. Fast-twitch, glycolytic type IIB fibres have a larger diameter and contract quickly. They produce high tension but for only a short time, as they fatigue quickly because of their anaerobic, lactic metabolism. Fast-twitch, oxidative-glycolytic type IIA fibres are intermediate between the other two, being moderately resistant to fatigue because of their mainly aerobic metabolism. These fibres are able to develop high tension but are susceptible to fatigue at high rates of activity.

The length–tension relationship

For a single muscle fibre, the tension developed when it is stimulated to contract depends on its length. Maximum tension occurs at about the resting length of the fibre, because the actin and myosin filaments overlap along their entire length, maximising the number of cross-bridges attached. If the fibre is stretched, the sarcomeres lengthen and the number of cross-bridges attached to the thin actin filaments decreases. Conversely, the shortening of the sarcomeres to below resting length results in the overlapping of actin filaments from opposite ends of the sarcomere. In both cases the active tension is reduced. In a whole muscle contraction, the passive tension caused by the stretching of the elastic elements must also be considered, as shown in Figure 6.12, as well as the active tension developed by the contractile component, which is similar to the active tension of an isolated fibre.

The total tension (Figure 6.12) is the sum of the active and passive tensions and depends upon the amount of connective tissue – the elastic elements – that the muscle contains. For single joint muscles, such as brachialis, the amount of stretch is not usually sufficient for the passive tension to be important. In two-joint muscles, such as three of the four heads of the hamstrings – semitendinosus, semimembranosus and the long head of biceps femoris – the extremes of the length–tension relationship may be reached, with maximal total tension being developed in the stretched muscle, as in Figure 6.12.

Figure 6.12 Length–tension relationship for whole muscle contraction.

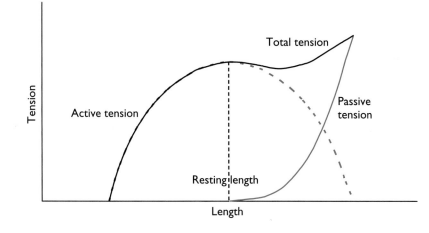

The force–velocity relationship

As Figure 6.13 shows, the speed at which a muscle shortens when concentrically contracting is inversely related to the external force applied; it is greatest when the applied force is zero. When the force has increased to a value equal to the maximum force that the muscle can exert, the speed of shortening becomes zero and the muscle is contracting isometrically. The reduction of contraction speed with applied force is accompanied by an increase in the latency period and a shortening of the contraction time. A further increase of the force results in an increase in muscle length as it contracts eccentrically and then the speed of lengthening increases with the force applied.

The tension–time relationship

The tension developed within a muscle is proportional to the contraction time. Tension increases with the contraction time up to the peak tension, as shown in Figure 6.14. Slower contraction enhances tension production as time is allowed for the internal tension produced by the contractile component, which can peak inside 10 ms, to be transmitted as external tension to the tendon through the series elastic elements; these have to be stretched, which may take about 300 ms. The tension within the tendon, and that transmitted to its attachments, reaches the maximum developed in the contractile component only if the duration of active contraction is sufficient. This only happens during prolonged tetanus.

Figure 6.13 Force–velocity relationship.

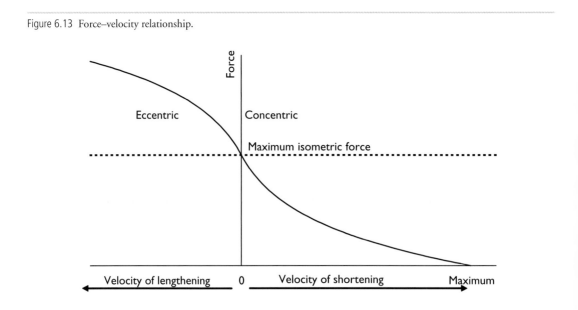

Figure 6.14 Tension–time relationship.

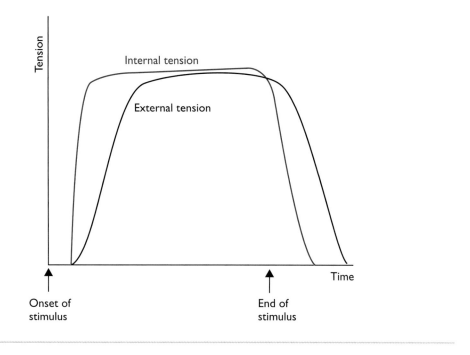

Muscle stiffness

The mechanical stiffness of a muscle is the instantaneous rate of change of force with length – it is the slope of the muscle tension–length curve. Unstimulated muscles possess low stiffness (or high compliance). This rises with time during tension and is directly related to the degree of filament overlap and cross-bridge attachment. At high rates of change of force, such as occur in many sports, muscle is stiff, particularly in eccentric contractions for which stiffnesses over 200 times those for concentric contractions have been reported. Stiffness is often considered to be under reflex control with regulation through both the length component of the muscle spindle receptors and the force–feedback component of the Golgi tendon organs. The exact role of the various reflex components in stiffness regulation in fast human movements in sport remains to be fully established as do their effects in the stretch–shortening cycle (see below). It is clear, however, that the reflexes can almost double the stiffness of the muscles alone at some joints. Furthermore, muscle and reflex properties and the central nervous system interact in determining how stiffness affects the control of movement.

Figure 6.15 Force potentiation in the stretch–shortening cycle in vertical jumps; blue curves show knee joint angle and black ones the vertical component of the ground reaction force: (a) concentric (+) knee extension in squat jump; (b) eccentric (–) contraction followed immediately by concentric (+) contraction in counter-movement jump; (c) as (b) but with a 1 s pause between the eccentric and concentric phases. The angles and forces have been 'normalised' equally across the three jumps to fall within the ranges 0 to +1 and –1 to +1, respectively.

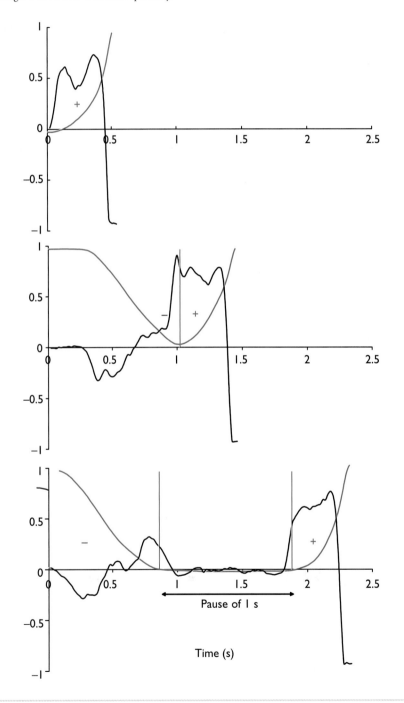

The stretch–shortening cycle

Many muscle contractions in dynamic movements in sport undergo a stretch–shortening cycle, in which the eccentric phase is considered to enhance performance in the concentric phase (Figure 6.15). The mechanisms thought to be involved are pre-load, elastic energy storage and release (mostly in tendon), and reflex potentiation. The stretch–shortening effect has not been accurately measured or fully explained. It is important not only in research but also in strength and power training for athletic activities. Some evidence shows that muscle fibres may shorten while the whole muscle–tendon unit lengthens. Furthermore, the velocity of recoil of the tendon during the shortening phase may be such that the velocity of the muscle fibres is less than that of the muscle–tendon unit. The result would be a shift to the right of the force–velocity curve (Figure 6.14) of the contractile component. These interactions between tendinous structures and muscle fibres may substantially affect elastic and reflex potentiation in the stretch–shortening cycle, whether or not they bring the muscle fibres closer to their optimal length and velocity. There have been alternative explanations for the phenomenon of the stretch–shortening cycle. Differences of opinion also exist on the amount of elastic energy that can be stored and its value in achieving maximal performance. The creation of larger muscle forces in, for example, a counter-movement jump compared with a squat jump is probably important both in terms of the pre-load effect and in increasing the elastic energy stored in tendon.

Muscle force components and the angle of pull

In general, the overall force exerted by a muscle on a bone can be resolved into three force components, as shown in Figure 6.16. These are:

Figure 6.16 Three-dimensional muscle force components: (a) side view; (b) force components; (c) front view.

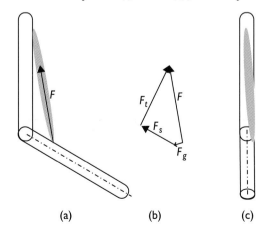

(a)   (b)   (c)

- A component of magnitude F_g, which tends to spin the bone about its longitudinal axis.
- A rotating component of magnitude F_t. In Figure 6.17 this is shown in one plane of movement only; in reality, this muscle force component may be capable of causing movement in both the sagittal and frontal planes as, for example, flexor carpi ulnaris flexing and adducting the wrist.
- A component of magnitude F_s along the longitudinal axis of the bone, which normally stabilises the joint, as in Figures 6.17(c) and (d). This component may sometimes tend to dislocate the joint, as in Figure 6.17(f). Joint stabilisation is an important function of muscle force, particularly for shunt muscles (see below).

The relative importance of the last two components of muscle force is determined by the angle of pull (θ); this is illustrated for the brachialis acting at the elbow in Figure 6.17(a), assuming F_s to be zero. The optimum angle of pull is 90° when $F_t = F$ (Figure 6.17(e)) and all the muscle force contributes to rotating the bone. The angle of pull is defined as the angle between the muscle's line of pull along the tendon and the mechanical axis of the bone. It is usually small at the start of the movement, as in Figure 6.17(b), and increases as the bone moves away from its reference position, as in Figures 6.17(e) and (f).

For collinear muscles, the ratio of the distance of the muscle's origin from a joint to the distance of its insertion from that same joint (a/b in Figure 6.17(b)) is known as the partition ratio, p. It can be used to define two kinematically different types of muscle. Spurt muscles are muscles for which $p > 1$; the origin is further from the joint than is the insertion. The muscle force mainly acts to rotate the moving bone. Examples are the biceps brachii and brachialis for forearm flexion. Such muscles are often prime movers. Shunt muscles, by contrast, have $p < 1$ because the origin is nearer the joint than is the insertion. Even for rotations well away from the reference position, the angle of pull is always small. The force is, therefore, directed mostly along the bone so that these muscles act mainly to provide a stabilising rather than a rotating force. An example is brachioradialis in forearm flexion. Such muscles may also provide the centripetal force, which is largely directed along the longitudinal axis of the bone towards the joint, for fast movements. Two-joint muscles are usually spurt muscles at one joint, as for the long head of biceps brachii acting at the elbow, and shunt muscles for the other joint, as for the long head of biceps brachii acting at the shoulder.

Within the human musculoskeletal system, anatomical pulleys serve to change the direction in which a force acts by applying it at a different angle and, sometimes, achieving an altered line of movement. The insertion tendon of peroneus longus is one such example. This muscle runs down the lateral aspect of the calf and passes around the lateral malleolus of the fibula to a notch in the cuboid bone of the foot. It then turns under the foot to insert into the medial cuneiform bone and the first metatarsal bone. The pulley action of the lateral malleolus and the cuboid accomplishes two changes of direction. The result is that contraction of this muscle plantar flexes the foot about the ankle joint (among other actions). Without the pulleys, the muscle would insert

Figure 6.17 Two-dimensional muscle force components: (a) brachialis acting on forearm and humerus; (b) angle of pull; (c) stabilising and turning components of muscle force; (d) to (f) effect of angle of pull on force components. The magnitude and direction of the muscle force vector (F) have been kept constant through figures (a) to (f).

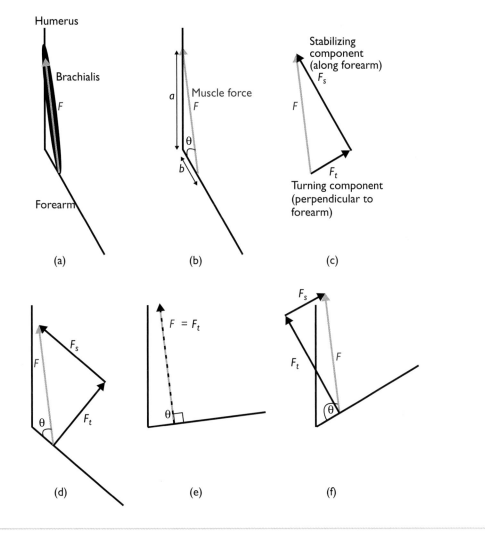

anterior to the ankle on the dorsal surface of the foot and would be a dorsiflexor. An anatomical pulley may also provide a greater angle of pull, thus increasing the turning component of the muscle force. The patella achieves this effect for quadriceps femoris, improving the effectiveness of this muscle as an extensor of the knee joint.

ELECTROMYOGRAPHY – WHAT MUSCLES DO

This and the next two sections are intended to provide you with an appreciation of the applications of electromyography to the study of muscle activity in sports movements. This includes the equipment and methods used and the processing of electromyographic data. We will also touch on the important relationship between muscle tension and the recorded signal, known as the electromyogram (abbreviation EMG, which is also, somewhat loosely, used as an abbreviation for electromyography). Electromyography is the technique for recording changes in the electrical potential of a muscle when it is caused to contract by a motor nerve impulse. The neural stimulation of the muscle fibre at the motor end-plate results in a reduction of the electrical potential of the cell and a spread of the action potential through the muscle fibre. The motor action potential (MAP), or muscle fibre action potential, is the name given to the waveform resulting from this depolarisation wave. This propagates in both directions along each muscle fibre from the motor end-plate before being followed by a repolarisation wave. The summation in space and time of motor action potentials from the fibres of a given motor unit is termed a motor unit action potential (MUAP, Figure 6.18). A sequence of MUAPs, resulting from repeated neural stimulation, is referred to as a motor unit action potential train (MUAPT). The physiological EMG signal is the sum, over space and time, of the MUAPT from the various motor units (Figure 6.18).

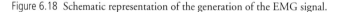

Figure 6.18 Schematic representation of the generation of the EMG signal.

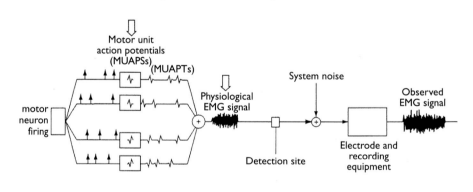

Electromyography is the only method of objectively assessing when a muscle is active. It has been used to establish the roles that muscles fulfil both individually and in group actions. The EMG provides information on the timing, or sequencing, of the activity of various muscles in sports movements. By studying the sequencing of muscle activation, the sports biomechanist can focus on several factors that relate to skill, such as any overlap of agonist and antagonist activity and the onset of antagonist activity at the end of a movement. It also allows the sports biomechanist to study changes in muscular activity during skill acquisition and as a result of training. Electromyography can also be used to validate assumptions about muscle activity that are made when calculating the internal forces in the human musculoskeletal system. It should, however, be noted that the EMG cannot necessarily reveal what a muscle is doing, particularly in fast multi-segment movements that predominate in sport.

Recording the myoelectric (EMG) signal

A full understanding of electromyography and its importance in sports biomechanics requires knowledge from the sciences of anatomy and neuromuscular physiology as well as consideration of many aspects of signal processing and recording, including instrumentation. A schematic representation of the generation of the (unseen) physiological EMG signal and its modification to the observed EMG is shown in Figure 6.18 – the EMG varies over time, as can be seen in this figure. So, we have another 'time series', which could be seen as another 'movement' pattern that could be useful to qualitative as well as quantitative movement analysts. Because of the complexity of the signal, it has rarely, if ever, been used in this way. Indeed, even quantitative biomechanical analysts often use it as no more than an 'on–off' indication of whether a muscle is contracting. As with kinematic data (Chapter 4), the EMG also consists of a range of frequencies, which can be represented by the EMG frequency spectrum (see below).

The physiological EMG signal is not the one recorded, as its characteristics are modified by the tissues through which it passes in reaching the electrodes used to detect it. These tissues act as a low-pass filter (Appendix 4.1, Chapter 4), rejecting some of the high-frequency components of the signal. The electrode-to-electrolyte interface acts as a high-pass filter, discarding some of the lower frequencies in the signal. The two-electrode, or bipolar, configuration, which is normal for sports electromyography, changes the two phases of the depolarisation–repolarisation wave to a three-phase one and removes some low-frequency and some high-frequency signals – it acts as a 'band pass' filter. Modern high-quality EMG amplifiers should not unduly affect the frequency spectrum of the signal but the recording device may also act as a band pass filter.

Many factors influence the recorded EMG signal. The intrinsic factors, over which the electromyographer has little control, can be classified as in Box 6.4.

BOX 6.4 INTRINSIC FACTORS THAT INFLUENCE THE EMG

- Physiological factors, which include the firing rates of the motor units; the type of fibre; the conduction velocity of the muscle fibres; and the characteristics of the volume of muscle from which the electrodes detect a signal (the detection volume) – such as its shape and electrical properties.
- Anatomical factors, which include the muscle fibre diameters and the positions of the fibres of a motor unit relative to the electrodes – the separation of individual MUAPs becomes increasingly difficult as the distance to the electrode increases.

Extrinsic factors, by contrast, can be controlled. These include the location of the electrodes with respect to the motor end-plates; the orientation of the electrodes with respect to the muscle fibres and the electrical characteristics of the recording system (see below). The use of equipment with appropriate characteristics is vitally important and will form the subject of the rest of this section. The electrical signals that are to be recorded are small – of the order of 10 μV to 5 mV. To provide a signal to drive any recording device, we require signal amplification. The general requirement is to detect the electrical signals (electrodes), modify the signal (amplifier) and store the resulting waveform (recorder) (compare with the force platform system of Chapter 5). All of this should be done linearly and without distortion. The following subsections summarise the important characteristics of EMG instrumentation.

EMG electrodes

The electrodes are the first link in the EMG recording chain. Their selection and placement are of considerable importance. Those used in sports biomechanics are predominantly surface electrodes, which can be passive, having no electrical power supply, and active, having a power supply. Indwelling electrodes are mainly used in clinical research. Few students have the luxury of choosing which type of electrode they will use and will normally use passive surface electrodes, which is the focus of the rest of these three EMG subsections. It is worth noting that compared with indwelling electrodes, surface electrodes are not only safer, easier to use and more acceptable to the person to whom they are attached but also, for superficial muscles, provide quantitative repeatability that compares favourably with indwelling ones. Many of the principles and procedures covered will also apply to active surface electrodes; other equipment and data processing are common to all types of electrode (see Burden, 2007; Further Reading, page 280).

The European Recommendations for Surface Electromyography (Hermens *et al.*, 1999; see Further Reading, page 280), which are called the SENIAM (Surface Electromyography for the Non-Invasive Assessment of Muscles) recommendations in the rest of this chapter, have attempted to standardise EMG assessment procedures. They recommend that only bipolar electrodes of pre-gelled silver–silver chloride material should be used, which conforms to standard practice in sports biomechanics. They also recommend a construction with a fixed distance of 20 mm between the centres of the two pre-mounted electrodes (see Figure 6.19). They found no objective reason to recommend any particular shape of electrode, instead advising that users should always report the type, manufacturer and shape of the electrodes used.

EMG cables

Electrical cables are needed to connect the EMG electrodes to the amplifier or pre-amplifier. These can cause problems, known as cable or movement artifacts, which

Figure 6.19 Bipolar configurations of surface electrodes.

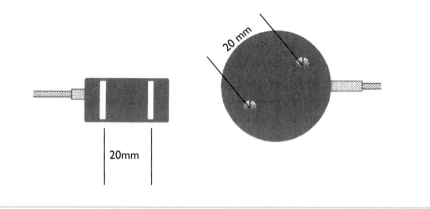

have frequencies in the range 0–10 Hz. The use of high-pass filtering, high-quality electrically shielded cables and careful taping to reduce cable movement can minimise these. Cable artifacts can be reduced by using pre-amplifiers mounted on the skin near the detection electrodes and by good experimental procedures (see next section). The use of active surface electrodes can virtually eliminate cable artifacts.

The effect of high-pass filtering at 10 Hz is shown in Figure 6.20, where Figures 6.20(a) and (b) are the unfiltered and filtered signals, respectively. The SENIAM recommendations for filtering the signal are to use a high-pass filter of 10 Hz if the signal is to be analysed in the frequency domain (see below) and 10–20 Hz if the signal is only to be used for movement analysis. To reduce high-frequency noise and aliasing, they also recommend a low-pass filter with a cut-off frequency of about 500 or 1000 Hz, at sampling frequencies, respectively, of 1000–2000 and 4000 Hz, depending on the application. Signal aliasing (see Chapter 4) will occur if the sampling rate is less than twice the upper frequency limit in the power spectrum of the sampled signal.

EMG amplifiers

These are the heart of an EMG recording system. They should provide linear amplification over the whole frequency and voltage range of the EMG signal. Noise must be minimised and interference from the electrical mains supply (mains hum) must be removed as far as possible. The input signal will be around 5 mV with surface electrodes (or 10 mV with indwelling electrodes). The most important amplifier characteristics are as follows.

Gain

This is the ratio of output voltage to input voltage and should be large and, ideally, variable in the range 100–10 000.

Figure 6.20 Effect of high-pass (10 Hz) filter on cable artifacts: (a) before filtering; (b) after filtering.

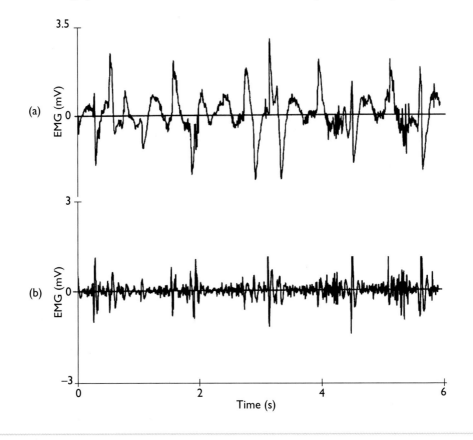

Input impedance

The importance of obtaining a low skin resistance can be minimised by using an amplifier with a high input impedance – by which we mean a high resistance to the EMG signal. This impedance should be at least 100 times the skin resistance to avoid attenuation of the input signal. This can be seen from the relationship between the ratio (e_i/e_{emg}) of the detected signal (e_{emg}) to the amplifier input signal (e_i), and the amplifier input impedance (R_i) and the two skin plus cable resistances (R_1 and R_2): $e_i/e_{emg} = R_i/(R_i + R_1 + R_2)$. The SENIAM recommendation for the minimum input impedance of an amplifier for use with passive surface electrodes is 100 MΩ (10^8 Ω). Input impedances for high-performance amplifiers – used with active electrodes – can be as high as 10 GΩ (10^{10} Ω).

Frequency response

The ability of the amplifier to reproduce the range of frequencies in the signal is known as its frequency response. The required frequency response depends upon the frequencies contained in the EMG signal. This is comparable with the requirement for audio amplifiers to reproduce the range of frequencies in the audible spectrum. Typical values of EMG frequency bandwidth are 10–1000 Hz for surface electrodes and 20–2000 Hz for indwelling electrodes. Most modern EMG amplifiers easily meet such bandwidth requirements. About 95% of the EMG signal is normally in the range up to 400 Hz, which explains why SENIAM recommend the use of a low-pass filter to remove higher frequencies. This signal range unfortunately also contains mains hum at 50 or 60 Hz. Also, SENIAM recommends that the input-referred voltage and current noise, respectively, should not exceed 1 μV and 10 pA.

Common mode rejection

Use of a single electrode would result in the generation of a two-phase depolarisation–repolarisation wave (Figure 6.21). It would also contain common mode mains hum. At approximately 100 mV, the hum is considerably larger than the EMG signal. This is overcome by recording the difference in potential between two adjacent electrodes, known as bipolar electrodes, using a differential amplifier. The hum is then largely eliminated as it is picked up commonly at each electrode because the body acts as an aerial – hence the name 'common mode'. The differential wave becomes triphasic (it has three phases, as in Figure 6.21). The smaller the electrode spacing, the more closely does the triphasic wave approximate to a time derivative of the single electrode wave. In practice, perfect elimination of mains hum is not possible and the success of its removal is expressed by the common mode rejection ratio (CMRR). This should be 10 000 or greater. The overall system CMRR can be reduced to a figure lower than that of the amplifier by any substantial difference between the two skin plus cable resistances. The effective system common mode rejection ratio in this case can be downgraded to a value of $R_1/(R_1 - R_2)$; cables longer than 1 m often exacerbate this problem. For passive electrodes, the attachment of the pre-amplifier to the skin near the electrode site reduces noise pick-up and minimises any degradation of the CMRR arising from differences between the cable resistances.

Recorders

Various devices have been used historically to record the amplified EMG signal but at present A–D conversion and computer processing are by far the most commonly used recording methods. An A–D converter with 12-bit (1 in 4096) or 16-bit resolution is recommended by SENIAM. High sampling rates are needed to reproduce successfully the signal in digital form; telemetered systems do not always provide a sufficiently high sampling rate.

Figure 6.21 EMG signals without mains hum; V_a and V_b are biphasic motor unit action potentials recorded from two monopolar electrodes, V_a–V_b is the triphasic differential signal from using the two electrodes in a bipolar configuration.

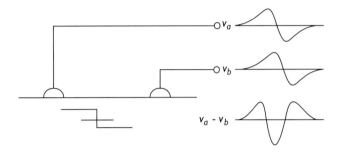

EMG and muscle tension

Obtaining a predictive relationship between muscle tension and the EMG could solve a major problem for quantitative sports biomechanists. This problem arises because the equations of motion at a joint cannot be solved because the number of unknown muscle forces exceeds the number of equations available. The problem is often known as muscle redundancy or muscle indeterminacy. If a solution could be found to this problem, it would allow the calculation of forces in soft tissue structures and between bones. It is not surprising, therefore, that the relationship between the EMG signal and the tension developed by a muscle has attracted the attention of many researchers.

The EMG provides a measure of the excitation of a muscle. Therefore, if the force in the muscle depends directly upon its excitation, a relationship should be expected between this muscle tension and suitably quantified EMG. A muscle's tension is regulated by varying the number and the firing rate of the active fibres; the amplitude of the EMG signal depends on the same two factors. It is, therefore, natural to speculate that a relationship does exist between EMG and muscle tension. We might further expect that this relationship would only apply to the active state of the contractile component, and that the contributions to muscle tension made by the series and parallel elastic elements would not be contained in the EMG (for a diagrammatical model of these elements of skeletal muscle, see Figure 6.10).

However, discrepancies exist between research studies even for isometric contractions, which should not lead us to expect a simple relationship between EMG and muscle tension for the fast, voluntary contractions that are characteristic of sports movements. For such movements, the relationship between EMG and muscle tension still remains elusive although the search for it continues to be worthwhile.

Additionally, in complex multi-segmental movements, such as those we observe in sport, muscles influence other joints as well as the ones they cross. This means that, in sports movements, the EMG tells us when a muscle is active but not, generally, what that muscle is doing. This is a very important limitation to the use of EMG.

EXPERIMENTAL PROCEDURES IN ELECTROMYOGRAPHY

The results of an electromyographic investigation can only be as good as the preparation of the electrode attachment sites. Although skin resistance ceases to be a problem if active electrodes are used, many EMG studies in sports biomechanics currently use passive surface electrodes, and good surface preparation is, therefore, still beneficial.

The location of the electrodes is the first consideration. For the muscles covered by their recommendations, the advice on electrode placements in the SENIAM report are well illustrated and simple (see Box 6.5 for a few examples). The separation of the electrodes was discussed above. For muscles not covered by those recommendations, simple advice is that:

- The two detector electrodes – or the detector pair – should be placed over the mid-point of the muscle belly.
- The orientation of the electrode pair should be on a line parallel to the direction of the muscle fibres (for example Box 6.5).
- If the muscle fibres are neither linear nor have a parallel arrangement, the line between the two electrodes should point to the origin and insertion of the muscle for consistency.

The following are good experimental practice for skin preparation to improve the electrode-to-skin contact, thereby reducing noise and artefacts:

- The area of the skin on which the electrodes are to be placed is shaved if necessary.
- The attachment area is then cleaned and degreased using an alcohol wipe, the alcohol being allowed to evaporate before the electrodes are attached to the skin. None of the procedures used historically, requiring abrasion with sandpaper or a lancet to scratch the skin surface, is recommended by SENIAM; such procedures raise ethical, or health and safety, issues at many institutions and are not necessary if the amplifier input impedance conforms to that recommended in the previous section.
- Preferentially, the electrodes should be pre-gelled; if they are not then, after the electrodes have been attached, electrode gel should be injected into the electrodes; any excess gel should be removed.
- It should not normally be necessary to check the skin resistance; however, if using older equipment, a simple DC ohmmeter reading of skin resistance should be taken. The resistance should be less than 10 kΩ and, preferably, less than 5 kΩ. If the resistance exceeds the former value, the electrodes should be removed and the preparation repeated.
- The reference electrode, also known as the earth or ground electrode, should be placed on electrically inactive tissue near the site of the recording electrodes or electrode pair. These sites are specific to each electrode (see examples in Box 6.5) and are usually on a bony prominence.

BOX 6.5 SOME ELECTRODE PLACEMENTS (ADAPTED FROM SENIAM)

Biceps brachii

Starting posture: Seated on a chair; elbow at a right angle; palm facing up.

Electrode location: On a line between the medial aspect of the acromion process and the cubit fossa at 1/3 of the distance from the latter (Figure 6.22(a)) in the direction of the muscle fibres.

Reference electrode: On or around the wrist.

Clinical test: With one hand under the elbow, flex the elbow slightly below, or at, a right angle with the forearm supinated, and press against the forearm to extend the elbow.

Triceps brachii – long head

Starting posture: Seated on a chair; shoulder about 90° abducted; elbow flexed at a right angle; palm facing down.

Electrode location: Midway along a line between the posterior crest of the acromion process and the olecranon process at two-finger widths medial to that line (Figure 6.22(b)) in the direction of the muscle fibres.

Reference electrode: On or around the wrist.

Clinical test: Extend the elbow while applying pressure that would tend to flex it.

Rectus femoris

Starting posture: Seated on a table; knees slightly flexed; upper body slightly bent backwards.

Electrode location: Midway along the line from the anterior superior iliac spine to the superior aspect of the patella (Figure 6.22(c)) in the direction of the muscle fibres.

Reference electrode: On or around the ankle or on the spinous process of the seventh cervical vertebra.

Clinical test: Extend the knee – with no rotation of the thigh – against pressure.

Biceps femoris

Starting posture: Lying prone with thighs on table; knees flexed by less than 90° from fully extended position; thigh and shank both slightly rotated externally.

Electrode location: Midway along the line from the ischial tuberosity to the lateral epicondyle of the tibia (Figure 6.22(d)) in the direction of that line.

Reference electrode: On or around the ankle or on the spinous process of the seventh cervical vertebra.

Clinical test: Extend the knee against pressure applied to the leg proximal to the ankle.

Figure 6.22 Electrode locations based on SENIAM recommendations: (a) biceps brachii; (b) triceps brachii; (c) rectus femoris; (d) biceps femoris.

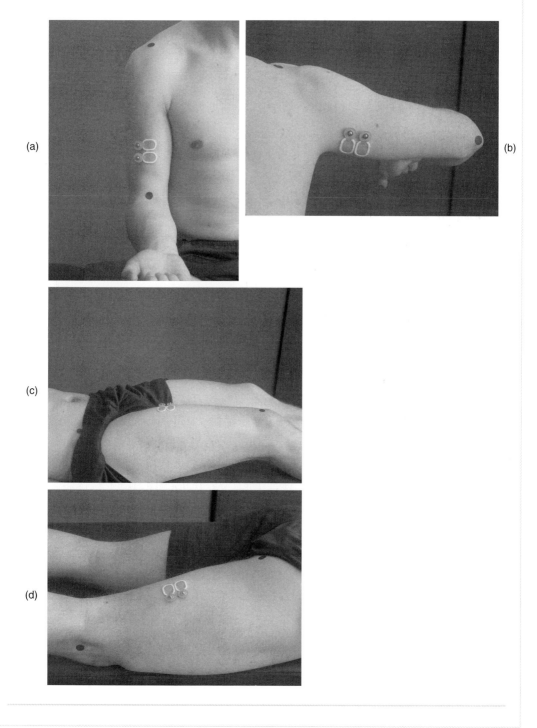

(a)

(b)

(c)

(d)

- After the electrodes have been attached to the skin, the electrodes and cables usually need to be fixed to prevent cable-movement artefacts and pulling of the cables; SENIAM recommends the use of rubber (elastic) bands or double-sided tape or rings for this purpose. They also recommend a clinical test for each individual muscle, performed from the starting posture, to ensure satisfactory signals; a few examples of these tests are shown in Box 6.5.

After use, the electrodes should be sterilised or disposed of.

EMG DATA PROCESSING

Much kinesiological, physiological and neurophysiological electromyography simply uses and analyses the raw EMG. The amplitude of this signal, and all other EMG data, should always be related back to the signal generated at the electrodes, not given after amplification. Further EMG signal processing is often performed in sports biomechanics in an attempt to make comparisons between studies. It can also assist in correlating the EMG signal with mechanical actions of the muscles or other biological signals.

Although EMG signal processing can provide additional information to that contained in the raw signal, care is needed for several reasons. First, to distinguish between artifacts and signal, it is essential that good recordings free from artifacts are obtained. Secondly, the repeatability of the results needs consideration; non-exact repetition of EMGs is likely even from a stereotyped activity such as treadmill running. The siting of electrodes, skin preparation and other factors can all affect the results. Even the activity or inactivity of one motor unit near the pick-up site can noticeably change the signal. Thirdly, such experimental factors make it difficult to compare EMG results with those of other studies. However, normalisation has been developed to facilitate such comparisons. This involves the expression of the amplitude of the EMG signal as a ratio to the amplitude of a contraction deemed to be maximal, usually a maximum voluntary contraction (MVC), from the same site. No consensus at present exists as to how to elicit an MVC and it is not always an appropriate maximum (see Burden, 2007, for elaboration of this point; Further Reading, page 280).

Temporal processing and amplitude estimation (time domain analysis)

Temporal processing relates to the amplitude of the signal content or the 'amount of activity'. It is often referred to as amplitude estimation. Such quantification is usually preceded by full-wave rectification as the raw EMG signal (Figure 6.23(a)) would have a mean value of about zero, because of its positive and negative deviations. Full-wave rectification (Figure 6.23(b)) simply involves making negative values positive; the rectified signal is expressed as the modulus of the raw EMG. Various terms have been,

Figure 6.23 Time domain processing of EMG: (a) raw signal; (b) full wave rectified (FWR) EMG; (c) smoothed, rectified EMG using a 6 Hz low-pass filter (linear envelope).

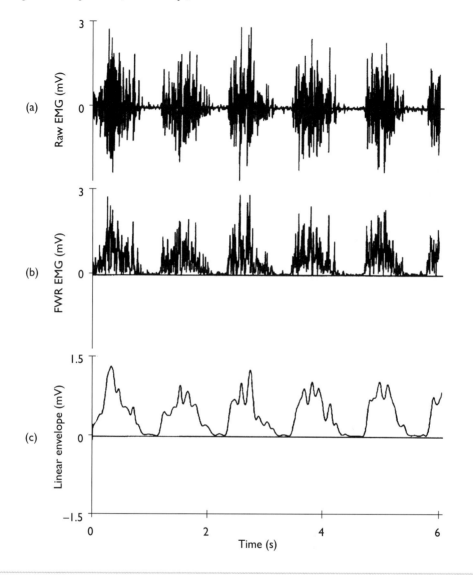

and are, used to express EMG amplitude, usually over a specified time interval, known as the epoch duration, such as the duration of the contraction or one running stride. The SENIAM recommendations specify epoch durations of 0.25–2 s for isometric contractions, 1–2 s for contractions less than 50% of an MVC and 0.25–0.5 s for contractions greater than 50% MVC. Such epoch durations considerably exceed the durations of muscle activity in many fast sports movements.

Average rectified EMG

The average rectified EMG is the average value of the full-wave rectified EMG, and is easily computed from a digital signal by adding the individual EMG values for each sample and dividing by the sample time. This is recommended by SENIAM for amplitude estimation of the EMG in non-dynamic contractions. The average rectified EMG is closely related to the integrated EMG (see below). As for other amplitude estimators of the EMG, the average rectified EMG is affected by the number of active motor units; the firing rates of motor units; the amount of signal cancellation by superposition; and the waveform of the MUAP. The last of these depends upon electrode position, muscle fibre conduction velocity, the geometry of the detecting electrode surfaces and the detection volume.

Root mean square EMG

The RMS EMG is the square root of the average power (voltage squared) of the signal in a given time. It is easily computed from a digital signal by adding the squares of the individual EMG values for each sample, then taking the square root of the sum before dividing by the sample time. It is considered to provide a measure of the number of recruited motor units during voluntary contractions where there is little correlation among motor units. This is also recommended by SENIAM for amplitude estimation of the EMG in non-dynamic contractions.

Integrated EMG

This is simply the area under the rectified EMG signal. It is not recommended at all by SENIAM and is mentioned here only because of its wide use – and abuse – in earlier research.

Smoothed, rectified EMG

The standard way of obtained the smooth, rectified EMG is by the use of a linear envelope detector (hence another name for this estimator, the 'linear envelope'), comprising a full-wave rectifier plus a low-pass filter. This is a simple method of quantifying signal intensity and gives an output (Figure 6.23(c)) that, it is claimed, follows the trend of the muscle tension curve with no high-frequency components. It is the only intensity detector recommended by SENIAM for dynamic contractions – which predominate in sport – because of the statistical properties of such contractions.

The main issue with this estimator is the choice of the epoch duration, which, for this estimator, is related to the filter cut-off frequency (see also Chapter 4). A low cut-off frequency gives a reliable estimate of the signal intensity for 'stationary' activation of the muscle – this is a statistical term meaning, roughly, that the statistical properties of the activation signal do not change with time, not that the activation is constant. However, the amplitude estimates with a low cut-off frequency are inaccurate if the activation is non-stationary, when the statistical properties of the signal vary with time. A high cut-off frequency results in a noisy estimator for stationary activations and a linear envelope that follows closely the changes in the EMG. Typical values of the cut-off frequency are around 2 Hz for slow movements such as walking, and around 6 Hz for faster movements such as running and jumping. Also, SENIAM recommends that the cut-off frequency used should always be reported along with the type and order of filter. Most software packages for EMG analysis in sports biomechanics use a standard fourth-order Butterworth filter (as for filtering kinematic data in Appendix 4.1).

A development of this approach is the use of an ensemble average calculated over several repetitions of the movement. This procedure theoretically – for physiologically identical movements – reduces the error in the amplitude estimation by a factor equal to the square root of the number of cycles. It is often used in clinical gait analysis but ignores the variability of movement patterns. It appears inappropriate for fast sports movements. The recommendations of SENIAM are that, if used, the number of cycles over which the ensemble average is calculated should be reported along with the standard error of the mean, to show up any movement variability.

Frequency domain analysis (spectral estimation)

Unless the data are explicitly time-limited, as in a single action potential, the EMG signal can be transformed into the frequency domain (as for kinematic data in Chapter 4). The EMG signal is then usually presented as a power spectrum (power equals amplitude squared) at a series of discrete frequencies, although it is often represented as a continuous curve, as in Figure 6.24. The epoch duration, over which the transformation of the time domain signal into the frequency domain occurs, which is recommended by SENIAM is 0.5–1.0 s. Again, these epoch durations considerably exceed the durations of muscle activity in many fast sports movements.

The central tendency and spread of the power spectrum are best expressed by the use of statistical parameters, which depend upon the distribution of the signal power over its constituent frequencies. Two statistical parameters are used to express the central tendency of the spectrum. These are comparable to the familiar mean and median in statistics. For a spectrum of discrete frequencies, the mean frequency is obtained by dividing the sum of the products of the power at each frequency and the frequency, by the sum of all of the powers. The median frequency is the frequency that divides the spectrum into two parts of equal power – the areas under the power spectrum to the two sides of the median frequency are equal. The median is less sensitive to noise than is the mean. The spread of the power spectrum can be expressed by the statistical bandwidth, which is calculated in the same way as a standard deviation.

The EMG power spectrum can be used, for example, to indicate the onset of muscle fatigue. This is accompanied by a noticeable shift in the power spectrum towards lower frequencies (Figure 6.25) and a reduction in the median and mean frequencies. This frequency shift is caused by an increase in the duration of the MUAP.

Figure 6.24 Idealised EMG power spectrum; f_m is the median and f_{av} the mean frequency.

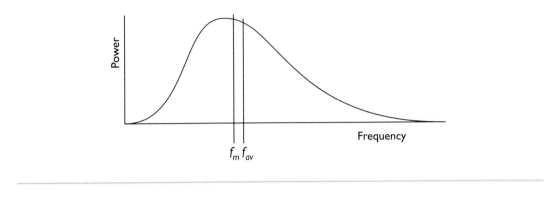

Figure 6.25 EMG power spectra at the start (a) and the end (b) of a sustained, constant force contraction.

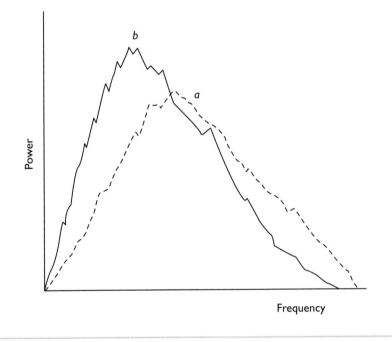

This results either from a lowering of the conduction velocity of all the action potentials or through faster, higher frequency motor units switching off while slower, lower frequency motor units remain active.

ISOKINETIC DYNAMOMETRY

The measurement of the net muscle torque at a joint using isokinetic dynamometry is very useful in providing an insight into muscle function and in obtaining muscle performance data for various modelling purposes ('isokinetic' is derived from Greek words meaning 'constant velocity'). Isokinetic dynamometry is used to measure the net muscle torque (called muscle torque in the rest of this section) during isolated joint movements, as in Figure 6.26. A variable resistive torque is applied to the limb segment under consideration; the limb moves at constant angular velocity once the preset velocity has been achieved, providing the person being measured is able to maintain that velocity in the specified range of movement. This allows the measurement of muscle torque as a function of joint angle and angular velocity, which, at certain joints, may then be related to the length and contraction velocity of a predominant prime mover, for example the quadriceps femoris in knee extension. By adjusting the resistive torque, both muscle strength and endurance can be evaluated. Isokinetic dynamometers are also used as training aids, although they do not replicate the types and speeds of movement in sport.

Passive isokinetic dynamometers operate using either electromechanical or hydraulic components. In these devices, resistance is developed only as a reaction to the applied muscle torque, and they can, therefore, only be used for concentric movements. Electromechanical dynamometers with active mechanisms allow for concentric and eccentric movements with constant angular velocity; some systems can be used for concentric and eccentric movements involving constant velocity, linearly changing acceleration or deceleration, or a combination of these.

Several problems affect the accuracy and validity of measurements of muscle torque using isokinetic dynamometers. Failure to compensate for gravitational force can result in significant errors in the measurement of muscle torque and data derived from those measurements. These errors can be avoided by the use of gravity compensation methods, which are an integral part of the experimental protocol in most computerised dynamometers. The development and maintenance of a preset angular velocity is another potential problem. In the initial period of the movement, the dynamometer is accelerated without resistance until the preset velocity is reached. The resistive mechanism is then activated and slows the limb down to the preset velocity. The duration of the acceleration period, and the magnitude of the resistive torque required to decelerate the limb, depend on the preset angular velocity and the athlete being evaluated. The dynamometer torque during this period is clearly not the same as the muscle torque accelerating the system. If the muscle torque during this period is required, it should be calculated from moment of inertia and angular acceleration data. The latter should be obtained either from differentiation of the position–time data or from accelerometers if these are available.

Errors can also arise in muscle torque measurements unless the axis of rotation of the dynamometer is aligned with the axis of rotation of the joint, estimated using anatomical landmarks (see, for example, Box 6.2). For normal individuals and small misalignments, the error is very small and can be neglected. Periodic calibration of the

Figure 6.26 Use of an isokinetic dynamometer for: (a) knee extension; (b) elbow flexion.

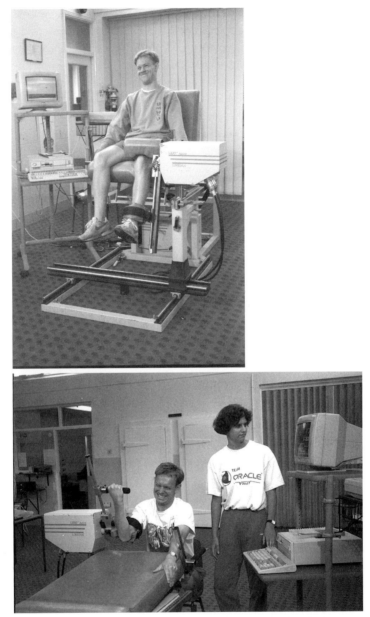

(a)

(b)

dynamometer system is necessary for both torque and angular position, the latter using an accurate goniometer. Torque calibration should be carried out statically, to avoid inertia effects, under gravitational loading.

Accurate assessment of isokinetic muscle function requires the measurement of torque output while the angular velocity is constant; computation of angular velocity is, therefore, essential. Most isokinetic dynamometers output torque and angular position data in digital form. Angular velocity and acceleration can be obtained by differentiation of the angular position-time data after using appropriate noise reduction techniques (see Chapter 4). Instantaneous joint power ($P = \boldsymbol{T}.\boldsymbol{\omega}$) can be calculated from the torque ($\boldsymbol{T}$) and angular velocity ($\boldsymbol{\omega}$) when the preset angular velocity has been reached.

Data processing in isokinetic dynamometry

The following parameters can normally be obtained from an isokinetic dynamometer to assess muscle function (for further information, see Baltzopoulos, 2007; Further Reading, page 280).

The maximum torque

The isokinetic maximum torque is used as an indicator of the muscle torque that can be applied in dynamic conditions. It is usually evaluated from two to six maximal repetitions and is taken as the maximum single torque measured during these repetitions. The maximum torque depends on the angular position of the joint. Maximum power can also be calculated.

The reciprocal muscle group ratio

The reciprocal muscle group ratio is an indicator of muscle strength balance around a joint, which is affected by age, biological sex and physical fitness. It is the ratio of the maximum torques recorded in antagonist movements, usually flexion and extension, for example the quadriceps femoris to hamstrings muscle group ratio.

The maximum torque position

The maximum torque position is the joint angular position at maximum torque and provides information about the mechanical properties of the activated muscle group. It is affected by the angular velocity. As the angular velocity increases, this position tends to occur later in the range of movement and not in the mechanically optimal joint position. It is, therefore, crucial to specify the maximum torque position as well as the maximum torque.

Muscular endurance under isokinetic conditions

Muscular endurance under isokinetic conditions is usually assessed through a 'fatigue index'. It provides an indication of the muscle group's ability to perform the movement at the preset angular velocity over time. Although there is no standardised testing protocol or period of testing, it has been suggested that 30–50 repetitions or a total duration of 30–60 s should be used. The fatigue index can then be expressed as the ratio of the maximum torques recorded in the initial and final periods of the test.

SUMMARY

In this chapter, we focused on the anatomical principles that relate to movement in sport and exercise. This included consideration of the planes and axes of movement and the principal movements in those planes. The functions of the skeleton, the types of bone, the process of bone fracture and typical surface features of bone were covered. We then looked at the tissue structures involved in the joints of the body, joint stability and mobility and the identification of the features and classes of synovial joints. The features and structure of skeletal muscles were considered along with the ways in which muscles are structurally and functionally classified, the types and mechanics of muscular contraction, how tension is produced in muscle and how the total force exerted by a muscle can be resolved into components depending on the angle of pull. The use of electromyography in the study of muscle activity in sports biomechanics was considered, including the equipment and methods used, and the processing of EMG data. Consideration was given to why the EMG is important in sports biomechanics and why the recorded EMG differs from the physiological EMG. We saw that electromyography shows us when a muscle is active but not, in complex multi-joint sports movements, what the muscle does. We covered the relevant recommendations of SENIAM and the equipment used in recording the EMG along with the main characteristics of an EMG amplifier. The processing of the raw EMG signal was considered in terms of its time domain descriptors and the EMG power spectrum and the measures used to define it. We concluded by examining how isokinetic dynamometry can be used to record the net muscle torque at a joint.

STUDY TASKS

Many of the following tasks can be attempted simply after reading the chapter. For some of the following study tasks, a skeleton, or a good picture of one, is helpful, while ones involving movements will require you to observe yourself or a partner. These observations will be easier if your experimental partner is dressed only in swimwear. If you are observing yourself, a mirror will be required.

1   With your experimental partner, perform the following activities for each of these synovial joints – the shoulder, the elbow, the radioulnar joints of the forearm, the wrist, the thumb carpometacarpal joint, the metacarpophalangeal and inter-phalangeal joints of the thumb and fingers, the hip, the knee, the ankle, the subtalar joint (rear foot), and the metatarsophalangeal and interphalangeal joints of the toes.

(a) Identify the joint's class and the number of axes of rotation (non-axial, uniaxial, biaxial or triaxial).

(b) Name and demonstrate all the movements at that joint.

(c) Estimate – from observation only – the range (in degrees) of each movement.

(d) Seek to identify the location of the axis of rotation for each movement and find a superficial anatomical landmark or landmarks (usually visible or palpable bony landmarks) that could be used to define this location – for example, you may find the flexion–extension axis of the shoulder to lie 5 cm inferior to the acromion process of the scapula.

Hint: You should reread the sections on 'The body's movements' (pages 225–32), 'The skeleton and its bones' (pages 232–7) and 'The joints of the body' (pages 237–41) before and while undertaking this task. You will also find information on the location of joint axes of rotation from anatomical landmarks in Box 6.2.

2   Have your experimental partner demonstrate the various movements of:

(a) The shoulder joint; observe carefully the accompanying movements of the shoulder girdle (scapula and clavicle) throughout the whole range of each movement

(b) The pelvis; observe the associated movements at the lumbosacral joint and the two hip joints.

Hint: You should reread the section on 'The body's movements' (pages 225–32) before undertaking this task.

3   Name the types of muscular contraction, demonstrating each in a weight-training activity, such as a biceps curl, and palpate the relevant musculature to ascertain which muscles are contracting.

Hint: You may wish to reread the subsections on 'Types of muscle contraction' (page 246) and 'Group action of muscles' (pages 246–7) and consult relevant material on muscle origins and insertions and prime mover roles of muscles on this book's website, or relevant sections in Marieb, 2003 (see Further Reading, page 280), before and while undertaking this task.

4   With reference to Figure 6.8 and using your experimental partner, locate and palpate all the superficial muscles in Figure 6.8. By movements against a light resistance only, seek to identify each muscle's prime mover roles.

Hint: You may wish to consult the material on muscle origins and insertions and prime mover roles of muscles on this book's website, or relevant sections in Marieb, 2003 (see Further Reading, page 280), before and while undertaking this task.

5   (a) How does the information contained in each of the time domain processed EMG signals differ from that in the raw EMG? What additional information might this provide for the sports biomechanist and what information might be lost?

(b) Outline the uses of the EMG power spectrum and the applications and limitations of the various measures used to describe it.

Hint: You may wish to consult the section on 'EMG data processing' (pages 268–72) before undertaking this task.

6 Explain the main characteristics required of an EMG amplifier. Log on to any EMG manufacturer's website (access through a search engine or through one of the web addresses on this book's website). Obtain the technical specification of that manufacturer's EMG amplifier, and ascertain whether this conforms to the recommendations of this chapter.
Hint: You may wish to consult the subsection on 'EMG amplifiers' (pages 261–3) before undertaking this task.

7 You are to conduct an experiment in which surface electrodes will be used to record muscle activity from biceps brachii, triceps brachii, rectus femoris and biceps femoris. Using a fellow student or friend to identify the muscles, and the recommendations of Box 6.5 for electrode placement, mark the sites at which you would place the detecting and ground electrodes for each of those muscles.
Hint: You may wish to consult the section on 'Experimental procedures in electromyography' (pages 265–8), as well as Box 6.5, before undertaking this task.

8 If you have access to EMG equipment, and with appropriate supervision if necessary, perform the preparation, electrode siting and so on from the previous exercise. Then carry out experiments to record EMGs as follows:

(a) From biceps brachii and the long head of triceps brachii during the raising and lowering phases of a biceps curl with a dumbbell and from an isometric contraction with the dumbbell held at an elbow angle of 90°. Check that you are obtaining good results and repeat the preparation if not. Comment on the results you obtain.

(b) From the same two muscles as in (a), throwing a dart – or similar object – at normal speed, much more slowly, and much more quickly. Comment on the results you obtain.

(c) From the rectus femoris and biceps femoris during rising from and lowering on to a chair. Again, check that you are obtaining good results, and repeat the preparation if not. Explain the apparently paradoxical nature of the results (this is known as Lombard's paradox).

Hint: You are strongly advised to consult, and follow under supervision, the section on 'Experimental procedures in electromyography' (pages 265–8), as well as Box 6.5, before and while undertaking this task. If you do not have access to EMG equipment, you can obtain EMGs from these experiments on the book's website.

You should also answer the multiple choice questions for Chapter 6 on the book's website.

GLOSSARY OF IMPORTANT TERMS (compiled by Dr Melanie Bussey)

Central tendency Measures of the location of the middle or the centre of a distribution.
Close-packed position The joint position with maximal contact between the articular surfaces and in which the ligaments are taut. See also **loose-packed position**.

Depolarisation A reduction in the potential of a membrane.

Differentiation Expresses the rate at which a variable changes with respect to the change in another variable on which it has a functional relationship.

Dynamometry The measurement of force or torque output; used as an estimate of muscular strength.

Elasticity The property of a material to return to its original size and shape. See also **plasticity** and **viscoelastic**.

Electrical potential The voltage across a membrane at steady-state conditions.

Frequency domain An analysis technique whereby the power of the signal is plotted as a function of the frequency of the signal. See also **power spectrum** and **time domain**.

Impedance (electrical) The ratio of voltage to electric current; a measure of opposition to time-varying electric current in an electric circuit. Sometimes identical to **resistance**.

Isokinetic exercise An exercise in which concentric muscle contraction moves a limb against a device that is speed controlled.

Isometric Muscle action in which tension develops but there is no visible or external change in joint position; no external work is produced. See also **isometric exercise**.

Isometric exercise An exercise that loads the muscle in one joint position.

Isotonic Muscle contraction in which tension is developed either by the lengthening (eccentric) or shortening (concentric) of muscle fibres. See also **isotonic exercise**.

Isotonic exercise An exercise in which an eccentric and or concentric muscle contraction is generated to move a specified weight through a range of motion.

Lag The number of data points by which a time series is shifted when one is calculating a cross-correlation with another time series or an autocorrelation with the same, but lagged, time series.

Loose-packed position The joint position with less than maximal contact between the articular surfaces and in which contact areas frequently change. See also **close-packed position**.

Maximal voluntary contraction The maximal force that is exerted by a muscle during a static contraction against an immovable resistance.

Motor end-plate A flattened expansion in the sarcolemma of a muscle that contains receptors to receive expansions from the axonal terminals; also called the **neuromuscular junction**.

Motor unit A motor neuron and all the muscle cells it stimulates.

Neuromuscular junction Region where the motor neuron comes into close contact with (innervates) skeletal muscle; also called the **motor end-plate**.

Normal stress The load per cross-sectional area applied perpendicular to the plane of cross-section of an object. See also **shear stress**.

Pelvic girdle The two hip bones plus sacrum, which can be rotated forwards, backwards and laterally to optimise positioning of the hip joint.

Plasticity Refers to the condition of connective tissue (ligaments or tendons) that has been stretched past its elastic limit and will no longer return to its original shape. See also **elasticity** and **viscoelastic**.

Polarisation Resting potential of a membrane.

Power spectrum The spectral density of a signal is a way of measuring the strength of the different frequencies that form the signal. See also **frequency domain**.

Reflex Involuntary response to a stimulus.

Resistance A measure of the extent to which an object, such as the skin, opposes the passage of an electric current. Sometimes identical to **impedance**.

Shear stress The load per cross-sectional area applied parallel to the plane of cross-section of the loaded object. See also **normal stress**.

Telemetry Automatic transmission and measurement of data from remote sources by radio or other remote means.

Tetanus State of muscle producing sustained maximal tension resulting from repetitive stimulation.

Time domain A variable that is presented as a function of time (a time series). See also **frequency domain**.

Viscoelastic A material that exhibits non-linear properties on a stress–strain curve. See also **elasticity** and **plasticity**.

FURTHER READING

Baltzopoulos, V. (2007) Isokinetic dynamometry, in C.J. Payton and R.M. Bartlett (eds) *Biomechanical Evaluation of Movement in Sport and Exercise*, Abingdon: Routledge. Chapter 6 provides a comprehensive and up-to-date coverage of all aspects of isokinetic dynamometry.

Basmajian J.V. and De Luca, C.J. (1985) *Muscles Alive: Their Functions Revealed by Electromyography*, Baltimore, MD: Williams and Wilkins. This is a classic text in its fifth edition, although the sixth edition is now far too long overdue. You should be able to find a copy in your university library. Chapters 12 to 17 provide a vivid description of the actions of muscles as revealed by electromyography, and are highly recommended. Other chapters cover, for example, motor control, fatigue and posture, but are now rather out of date.

Burden, A.M. (2007) Surface electromyography, in C.J. Payton and R.M. Bartlett (eds) *Biomechanical Evaluation of Movement in Sport and Exercise*, Abingdon: Routledge. Chapter 5 provides an up-to-date coverage of many aspects of electromyography related to sports movements.

Hermens, H.J., Freriks, B., Merletti, R., Stegeman, D., Blok, J., Rau, G., Disselhorst-Klug., C. and Hägg, G. (1999) *SENIAM: European Recommendations for Surface Electromyography*, Entschede: Roessingh Research and Development. The most recent attempt to standardise procedures for electromyography. A very valuable reference source although it is spoilt somewhat by typographical, punctuation and grammatical errors and is rather turgid. There are also some bizarre omissions of superficial muscle placement sites, such as those for all the muscles that originate in the forearm; however, the placement sites recommended are very useful for both students and researchers.

Marieb, F.N. (2003) *Human Anatomy and Physiology*, Redwood City, CA: Benjamin/Cummings. See Chapters 6 to 10. Many anatomy and physiology texts will contain supplementary information about, for example, the attachment points and actions of specific muscles. This one is a highly recommended and readable text with glorious colour illustrations.

Index